Contents

Introduction

About this guide

This unit guide covers essential material for the OCR AS Physics A specification. It contains material needed for the effective study of **Unit G482: Electrons, Waves and Photons**. This unit is assessed by a single written examination lasting 1 hour 45 minutes. The aim of the guide is to help you *understand* the physics, so that you can effectively revise and prepare for the examination.

The **Content Guidance** section is based on the structure of the specification. The same headings are used and the sub-headings closely follow the learning outcomes of the specification. This guide is not meant to be a detailed textbook and does not contain every fact that you need to know. The focus is on understanding the principles and definitions so that you can tackle successfully the variety of questions in the examination. This guide uses worked examples to illustrate good practice and offers an examiner's perspective on how to improve your answers.

The **Question and Answer** section shows the type of questions you can expect in the unit examination. The questions are illustrated with answers given by a typical C-grade candidate and an A-grade candidate. The answers are followed by comments from the examiner explaining why the marks were awarded or lost. Common errors made by candidates are also highlighted, so that you do not make the same mistakes. Candidates frequently lose marks for careless mistakes, incomplete answers, muddled presentation and illegible handwriting.

Use of this guide

This guide can be used throughout the course and not just at revision time. The content of the guide is set out in the order of the learning outcomes of the specification, so that you can use it:

- to check your notes
- as a reference for homework and internal tests
- to revise in manageable sections
- to check that you have covered the specification completely
- to identify your strengths and weaknesses
- to identify terms used by examiners
- to familiarise yourself with the range of questions you can expect in the exam
- to improve the quality of your answers
- to increase your confidence in the application of physics

Revision

Planning study and revision is essential. You can avoid disappointment and anxiety by organising a revision plan well before any pending examination. You cannot learn the content of the whole course in a couple of days. Some points to consider when revising for an examination include:

- Always start with a topic that you find easy. This will boost your confidence.
- Learn equations, definitions and laws in the specification thoroughly. Examiners expect perfection with definitions and laws and tend to give no marks for using wrong equations.
- Ask your teacher to explain any words, laws or definitions that you do not understand.
- Make your revision active by writing out the equations, definitions and laws, drawing diagrams and doing calculations.
- Write a brief summary of the topic.
- When revising, make sure you have your notes, the specification and a reference book available to complement this guide.
- Make good use of the specification. Use a highlighter pen to identify the topics that you have covered.
- Do not revise for long periods. If you are tired, you cannot produce quality work.
- Do not leave your revision to the last moment. Plan out a strategy spread over several weeks before the examination. Work hard during the day and learn to relax when needed.

Key skills

In examinations, candidates lose many marks because of their inability to apply some basic skills.

Calculator work

Some common mistakes are highlighted below.

- $\frac{3+4}{0.5}$ is put into the calculator as [3] [+] [4] [÷] [0.5], which gives 11. It should be [(] [3] [+] [4] [)] [÷] [0.5], which gives the correct answer of 14.
- $\frac{4}{10\times2}$ is put into the calculator as [4] [÷] [10] [×] [2], which gives 0.8. It should be [4] [÷] [(] [10] [×] [2] [)] or [4] [÷] [10] [÷] [2], which gives the correct answer of 0.2.
- 3.0×10^8 is put into the calculator as [3.0] [×] [10] [EXP] [8], which gives 3.0×10^9. It should be [3.0] [EXP] [8]. (On some calculators, the 'exponent' button is 10^x.)
- The minus sign is not inserted properly for powers of 10.
- Squares are not put in when they are needed.

Numerical work

Some common mistakes are highlighted below.

- An inappropriate number of significant figures is used. If the data in a question are given to two significant figures, your answer must also be given to two significant figures. Examiners tend not to penalise an answer with too many significant figures. Candidates lose more marks by using too few significant figures.
- Rounding up during a long calculation can lead to an incorrect answer. It is best to keep all the significant figures on your calculator as you progress through a calculation.
- Significant figures and decimal places may be confused. The number 0.0254 is three significant figures, written to four decimal places. Using standard form, 2.54×10^{-2} is clearly written to three significant figures.
- Answers are not checked to see whether they are sensible. For example, the mass of a car is determined to be –600 kg. You should realise that a negative mass is ridiculous. This negative answer should act as a prompt to check for an error in the calculation.

Algebraic work

Candidates lose marks by not being able to complete some simple algebraic operations. Here are some rules worth knowing:

- $a = b + c$ hence, $b = a - c$
- $a = bc$ hence, $b = \frac{a}{c}$
- $a = \frac{b}{c}$ hence, $b = ac$ and $c = \frac{b}{a}$
- $a^2 = b$ hence, $a = \sqrt{b}$

Descriptive work

In examinations, candidates tend to gain more marks from mathematical questions than questions requiring descriptive answers. Descriptive questions are often answered badly for the following reasons:

- The answer is of an inappropriate length. You need to be aware of the number of marks available and write accordingly. Answers that are too long waste time and may be repetitive; answers that are too short almost inevitably miss out key points that the examiners are looking for.
- There is a failure to state and use physics principles and vocabulary in answers that require extended writing.
- The structure of the written answer is poor. The answer rambles with candidates writing down ideas as and when they think of them. Think carefully about the physics required.
- Poor sentence construction and bad spelling lose marks for **q**uality of **w**ritten **c**ommunication (QWC).
- The question is misinterpreted. Read the question carefully and start writing the answer only once you are sure that you have understood the question.

Answering examination questions

Examiners do not set questions to trick you. They simply want you to demonstrate your knowledge and understanding of physics. Examiners spend many hours discussing the wording of questions and their aim is to give you the necessary information succinctly, so that you do not have to waste time deciphering the question.

When answering a question, you are expected to present your ideas logically. In an extended writing question, present your ideas in clear steps that show good use of your knowledge of physics. In a calculation, you should show clearly the following:

- correct equation
- correct substitution in the right units
- correct algebraic manipulation
- correct answer with the units

Do not waste your time in the examination. You are expected on average to secure 1 mark each minute. Read the question carefully before you put pen to paper. Make sure that all calculations are done in the correct units and that you have taken into account prefixes such as 'milli-' and 'kilo-'. Check your work as you go along. It is not sensible to check all your answers at the end of the paper because you will have forgotten the finer details of each question. You can, however, do a quick check for units and significant figures once you have finished the paper and have some spare time at the end of the session.

Candidates waste too much time drawing diagrams using rulers. In an examination, most diagrams can be drawn freehand. There is no point in drawing a circuit diagram with the skills of a draughtsman when a freehand sketch showing all the components will do. Learn to save time in an examination.

In descriptive answers, do not use bullet points if marks are available for quality of written communication.

In the Question and Answer section of this guide, there are comments on the mistakes that candidates make. In the run-up to an exam, make sure you read the examiner's comments at the end of each question.

Command words used in examinations

The list below shows the most frequently used command words and their meanings:

- **Calculate** — this is used when a numerical answer is required. Show all your working and give an appropriate unit for your final answer. The number of significant figures must reflect the given data.
- **Deduce** — you have to draw conclusions from the information provided.
- **Define** — a formal statement or a word equation is required. Do not use symbols.
- **Describe** — this requires you to state in words, and diagrams if appropriate, the main points of the topic. The amount of description depends on the mark allocation.

- **Determine** — this often implies that the quantity cannot be determined directly but can be deduced from the information available.
- **Estimate** — this requires a statement or calculation in which you make sensible assumptions and use realistic values for quantities.
- **Explain** — you have to use correct physics vocabulary and principles. Normally a definition should be given, together with a relevant comment. The depth of the answer depends on the mark allocation.
- **Select** — you will be given a list of key equations. Make sure that you choose the correct equation for the calculation.
- **Show** — the answer to a particular problem is given on the question paper and is required for a subsequent calculation. You need to show every stage of your working. This is not a calculator exercise to find ways of arriving at the correct answer.
- **Sketch** — a simple freehand drawing is required. Significant detail should be included.
- **Sketch a graph** — the shape of the graph should be qualitatively correct. Values of the intercept or gradient may be required. Axes must be labelled and the origin shown if this is appropriate.
- **State** — this implies a brief answer with little, if any, supporting argument.
- **Suggest** — there is often no single correct answer. You will be given credit for sensible reasoning based on correct physics.

Data, formulae and relationships

You will be given the following information when you take the examination for Unit G482: Electrons, Waves and Photons.

Data

Values are given to three significant figures, except where more are useful.

Speed of light in a vacuum	c	$3.00 \times 10^{8}\ \mathrm{m\,s^{-1}}$
Permittivity of free space	ε_0	$8.85 \times 10^{-12}\ \mathrm{C^2\,N^{-1}\,m^{-2}\ (F\,m^{-1})}$
Elementary charge	e	$1.60 \times 10^{-19}\ \mathrm{C}$
Planck constant	h	$6.63 \times 10^{-34}\ \mathrm{J\,s}$
Gravitational constant	G	$6.67 \times 10^{-11}\ \mathrm{N\,m^2\,kg^{-2}}$
Avogadro constant	N_A	$6.02 \times 10^{23}\ \mathrm{mol^{-1}}$
Molar gas constant	R	$8.31\ \mathrm{J\,mol^{-1}\,K^{-1}}$
Boltzmann constant	k	$1.38 \times 10^{-23}\ \mathrm{J\,K^{-1}}$
Electron rest mass	m_e	$9.11 \times 10^{-31}\ \mathrm{kg}$
Proton rest mass	m_p	$1.673 \times 10^{-27}\ \mathrm{kg}$
Neutron rest mass	m_n	$1.675 \times 10^{-27}\ \mathrm{kg}$
Alpha particle rest mass	m_α	$6.646 \times 10^{-27}\ \mathrm{kg}$
Acceleration of free fall	g	$9.81\ \mathrm{m\,s^{-2}}$

Conversion factors

Unified atomic mass unit	$1\ u = 1.661 \times 10^{-27}\ kg$
Electronvolt	$1\ eV = 1.60 \times 10^{-19}\ J$
Time	$1\ day = 8.64 \times 10^{4}\ s$
	$1\ year \approx 3.16 \times 10^{7}\ s$
	$1\ light\ year \approx 9.5 \times 10^{15}\ m$

Mathematical equations

arc length = $r\theta$

circumference of circle = $2\pi r$

area of circle = πr^2

curved surface area of cylinder = $2\pi rh$

volume of cylinder = $\pi r^2 h$

surface area of sphere = $4\pi r^2$

volume of sphere = $\frac{4}{3}\pi r^3$

Pythagoras' theorem: $a^2 = b^2 + c^2$

For small angle $\theta \Rightarrow \sin\theta \approx \tan\theta \approx \theta$ and $\cos\theta \approx 1$

$\lg(AB) = \lg(A) + \lg(B)$

$\lg\left(\frac{A}{B}\right) = \lg(A) - \lg(B)$

$\ln(x^n) = n\ln(x)$

$\ln(e^{kx}) = kx$

Formulae and relationships

Listed below are the formulae and relationships that you are given for Unit G482: Electrons, Waves and Photons.

$\Delta Q = I\Delta t$

$I = Anev$

$W = VQ$

$V = IR$

$R = \frac{\rho L}{A}$

$P = VI \quad P = I^2R \quad P = \frac{V^2}{R}$

$W = VIt$

e.m.f. = $V + Ir$

$$V_{out} = \frac{R_2}{R_1 + R_2} \times V_{in}$$

$$v = f\lambda$$

$$\lambda = \frac{ax}{D}$$

$$d\sin\theta = n\lambda$$

$$E = hf \quad E = \frac{hc}{\lambda}$$

$$hf = \phi + \mathrm{KE}_{max}$$

$$\lambda = \frac{h}{mv}$$

$$R = R_1 + R_2 + \ldots$$

$$\frac{1}{R} = \frac{1}{R_1} + \frac{1}{R_2} + \ldots$$

Content Guidance

The content guidance section is a student's guide to AS Unit G482: Electrons, Waves and Photons. The main topics are:

- Electric current
- Circuit symbols
- Electromotive force (e.m.f.) and potential difference (p.d.)
- Resistance
- Resistivity
- Power
- Series and parallel circuits
- Practical circuits
- Wave motion
- Electromagnetic waves
- Interference
- Stationary waves
- Energy of a photon
- The photoelectric effect
- Wave–particle duality
- Energy levels in atoms

This section covers all the relevant key facts, explains the essential concepts and highlights some common misconceptions.

Electric current

Current and charge

An electric current in a circuit is caused by the movement of **charged particles**.

In a metal, the current is due to the movement of **electrons**. Electrons are negatively charged. The magnitude of the charge on an electron is known as the elementary charge *e*.

The experimental value for e is 1.6×10^{-19} C.

In a conducting solution (electrolyte), the current is due to the simultaneous movement, in opposite directions, of positive and negative **ions**.

The diagram below shows two simple circuits:

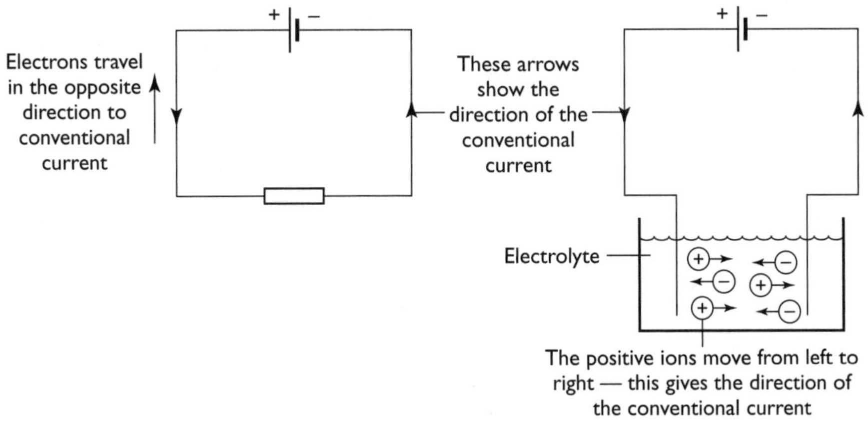
+
−
Electrons travel in the opposite direction to conventional current
These arrows show the direction of the conventional current
+
−
Electrolyte
The positive ions move from left to right — this gives the direction of the conventional current

The convention for the direction of current in a circuit is from the positive (+) terminal of the cell to the negative (−) terminal of the cell. This convention was agreed long before the electron was discovered. The **conventional current** is in the *opposite* direction to the flow of electrons.

Electric current is measured in **amperes** (A) and charge is measured in **coulombs** (C).

Electric current is the rate of flow of charge:

$$\textbf{current} = \frac{\textbf{charge flow}}{\textbf{time}}$$

or

$$I = \frac{\Delta Q}{\Delta t}$$

where I is the current at a point in a circuit, ΔQ is the charge flow past the point in a time interval Δt. The symbol Δ (delta) is shorthand for 'change in'. From the equation above it follows that:

1 ampere = 1 coulomb per second or $\mathbf{1\,A = 1\,C\,s^{-1}}$

Worked example

The current in a wire is 64 mA. Calculate:

(a) the charge passing through a point in the wire in a time of 20 s

(b) the number of electrons responsible for the charge in **(a)**

Answer

(a) $I = \frac{\Delta Q}{\Delta t}$

Therefore, $\Delta Q = I\Delta t = 0.064 \times 20$

$\Delta Q = 1.28\,\text{C}$

The charge flow is 1.28 C.

(b) charge = number of electrons × e

number of electrons = $\frac{\Delta Q}{e} = \frac{1.28}{1.6 \times 10^{-19}}$

number of electrons = 8.0×10^{18}

Tip The number of electrons is 8 000 000 000 000 000 000. You do not want an examiner to have to count all the zeros — hence it is convenient to write the answer in standard form.

Current against time graph

The diagram below shows a current against time graph for current at a point in a circuit.

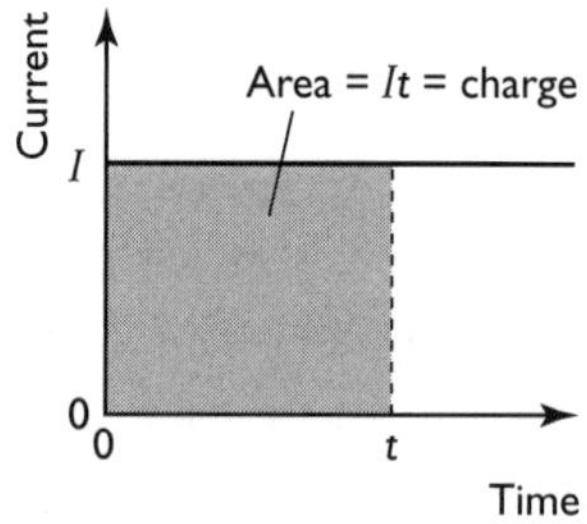

The current I is constant. The charge flow in a time t is given by:

charge = It

The area under the graph is also equal to the charge flow. In fact, the area under a current against time graph is *always* equal to the charge flow. It does not depend on the shape of the graph. Therefore:

area under a current against time graph = charge flow

Kirchhoff's first law

Kirchhoff's first law states that:

The sum of the currents entering a point (or junction) in a circuit is equal to the sum of the currents leaving the same point.

This law is an expression of the **conservation of charge** — for example, if a million electrons enter a point in a circuit, then the same number of electrons must exit that point. Hence, the same amount of charge must enter and exit a point in a given time. The diagrams below illustrate the law for a number of situations. Notice that the sum of currents entering the point **X** is equal to the total current exiting from this point.

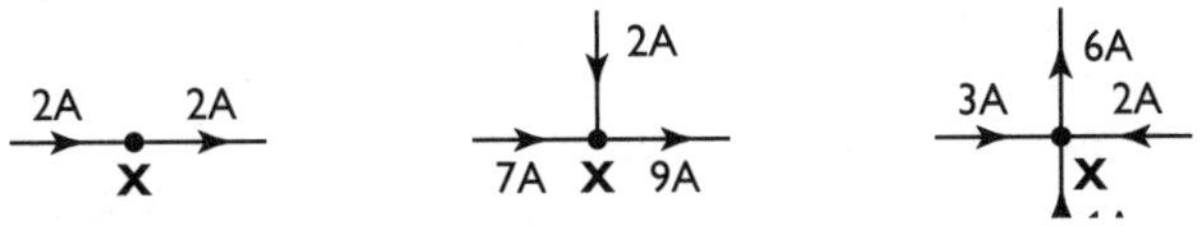

Mean drift velocity

When a metal wire is connected to a cell, the current in the wire is caused by the movement of electrons. The electrons experience an electrical force due to the terminals of the cell. This force accelerates the electrons until they collide with the vibrating metal ions. Between collisions with the ions, the speed of the electrons can be as high as $10^6\,m\,s^{-1}$. The journey of an electron towards the positive terminal of the cell is haphazard because of the random collisions of the electrons with the ions. The **mean drift velocity** of the electrons along the length of the wire may be only a fraction of a millimetre per second ($10^{-5}\,m\,s^{-1}$).

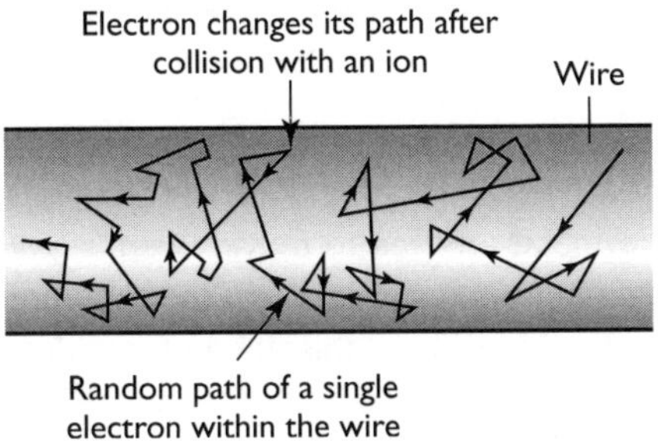

Tip Do not suggest that the collisions are between the electrons themselves. This is a common mistake made by candidates.

The magnitude of the current I in a material is given by the equation:

$$I = Anev$$

where A is the cross-sectional area of the material, n is the number of electrons per unit volume, e is the elementary charge and v is the mean drift velocity of the electrons.

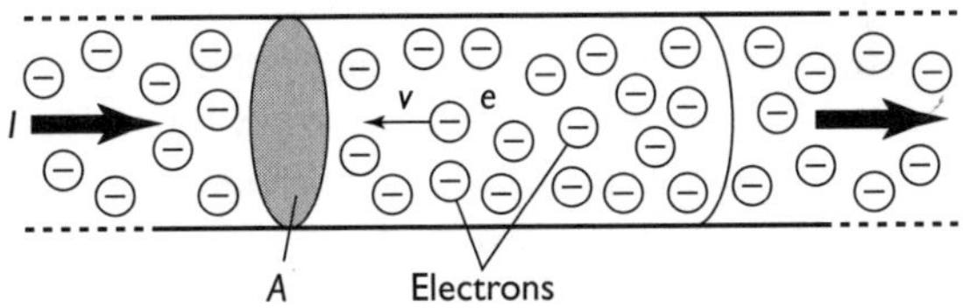

The number of electrons per unit volume, n, is also known as the **number density** of the electrons.

Worked example

A copper cable carrying current to an electric kettle has cross-sectional area of $3.6 \times 10^{-6}\,m^2$. It delivers a constant current of 8.0 A for a period of 3.0 minutes. The number density of electrons for copper is $8.5 \times 10^{28}\,m^{-3}$. Calculate:

(a) the mean drift velocity of the electrons

(b) the average distance travelled by an electron along the length of the cable in a time of 3.0 minutes

Answer

(a) $I = Anev$

Therefore, $v = \frac{I}{Ane}$

$$v = \frac{8.0}{3.6\times10^{-6}\times8.5\times10^{28}\times1.6\times10^{-19}}$$

$v = 1.63 \times 10^{-4}\,m\,s^{-1} \approx 0.16\,mm\,s^{-1}$

(b) distance = speed × time

distance = $1.63 \times 10^{-4} \times (3.0 \times 60)$

distance = 0.029 m (2.9 cm)

Tip Do not forget to change the time from minutes into seconds.

Metals, semiconductors and insulators

Not all the electrons in a metal are free to move. Even in good electrical conductors such as copper and aluminium, only about 3% of the total number of electrons in the metal take part in conduction. In spite of this, the number density of metals is approximately $10^{28}\,m^{-3}$.

Insulators (e.g. plastics) are poor electrical conductors because they have fewer free electrons per unit volume. Conduction in a semiconductor such as silicon is quite complex. However, it is enough to mention that the number density of the electrons lies between those of metals and insulators. The number density of electrons in silicon is about $10^{18}\,m^{-3}$, which is approximately 10 000 000 000 times smaller than that in a metal.

Circuit symbols

Make sure that you are familiar with all the circuit symbols mentioned in the specification. Take particular care in examinations when drawing these symbols. Too many candidates lose vital marks for sloppy drawing. The diagram on p. 17 shows the symbols used most frequently in Unit G482 exams.

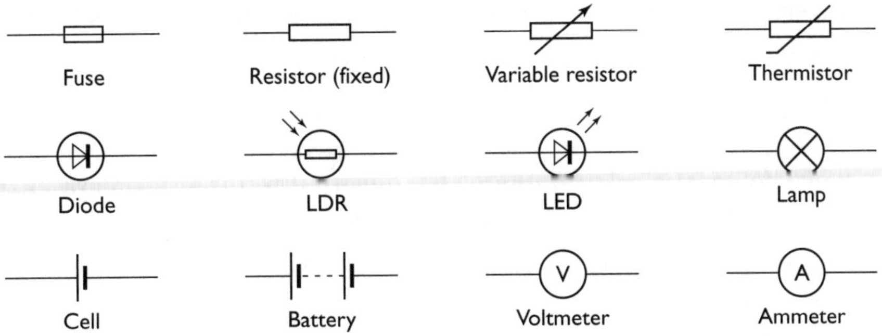

Electromotive force (e.m.f.) and potential difference (p.d.)

Energy transformation in a circuit

The circuit diagram below shows two components connected in series to a cell.

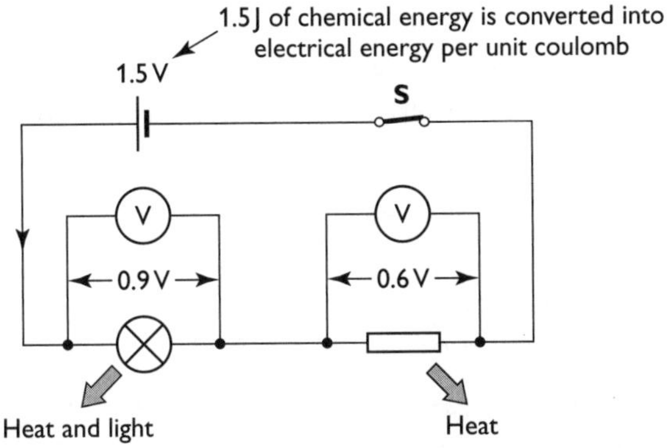

There is a current in the circuit when the switch **S** is closed. Electrons travelling through the cell *gain* electrical energy. Within the cell, chemical energy is transformed into electrical energy. As the electrons travel through the external components, they *lose* this electrical energy as heat and light in the lamp and as heat in the resistor. The voltmeter placed across each component measures the **potential difference** (p.d.) or **voltage** across the component.

Potential difference across a component is defined as the energy lost by the charges per unit charge:

$$\textbf{potential difference} = \frac{\textbf{energy lost by charges}}{\textbf{charge flow}}$$

$$V = \frac{W}{Q}$$

where V is the potential difference across the component, W is the amount of energy lost as heat (and light) and Q is the charge flow through the component.

Potential difference is measured in **volts** (V). From the definition above, it follows that:

1 volt = 1 joule per coulomb or $\mathbf{1\,V = 1\,J\,C^{-1}}$

The unit of potential difference, the volt, is defined as follows:

A potential difference of 1 volt exists between two points in a circuit when 1 joule of electrical energy is converted to other forms of energy per unit charge travelling between the points.

In the circuit diagram on p. 17, the voltmeter measures 0.9 V across the lamp and 0.6 V across the resistor. Hence:

- 0.9 J of electrical energy is converted to heat and light per coulomb of charge passing through the lamp
- 0.6 J of electrical energy is converted into heat per coulomb passing through the resistor

Tip It would be sensible for you to think of a voltmeter as a device that measures 'joules per coulomb'.

Without the cell, there is no movement of electrons and hence there is no transfer of electrical energy to other forms round the circuit. The energy transferred in the external components must originally have come from the cell. The principle of conservation of energy implies that the cell converts 1.5 J of chemical energy into electrical energy per unit charge passing through it. The electromotive force (e.m.f.) of the cell is $1.5\,J\,C^{-1}$ or, simply, 1.5 V.

The electromotive force of the cell is defined as the energy *gained* by charges per unit charge:

$$\textbf{e.m.f.} = \frac{\textbf{energy gained by charges}}{\textbf{charge flow}} \quad or\ E = \frac{W}{Q}$$

where E is the electromotive force of the cell, W is the amount of energy gained as electrical energy by the charges and Q is the charge flow through the cell. The letter W is used for energy in both definitions. This seems unusual at first, but do not forget that 'energy transfer = *w*ork done'.

The key features of electromotive force and potential difference (voltage) are summarised in the table below.

	e.m.f.	p.d.
Measured in volts (V)	✓	✓
Defined as $\frac{\text{energy}}{\text{charge}}$	✓	✓
Charges gain electrical energy	✓	
Charges lose electrical energy into other forms (e.g. heat)		✓

All sources of electromotive force transform one form of energy into electrical energy. Here are some examples:

- solar cell — transforms light energy into electrical energy
- thermocouple — transforms heat energy into electrical energy
- dynamo — transforms kinetic energy into electrical energy

Worked example

A chemical cell transforms 6.0 J of chemical energy into electrical energy when a charge of 4.0 C passes through it.

(a) What is the electromotive force of the cell?

(b) The cell is connected to two identical lamps in series. Determine the potential difference across each lamp.

Answer

(a) $\text{e.m.f.} = \dfrac{\text{energy}}{\text{charge}}$

$\text{e.m.f.} = \dfrac{6.0}{4.0} = 1.5\,\text{V}$

The electromotive force of the cell is 1.5 V.

(b) The 4.0 C charge moving round the circuit loses 6.0 J of energy as heat and light in the two lamps. Since the lamps are identical, the energy lost by the charges in each lamp is 3.0 J. Therefore the potential difference across each lamp is:

$\text{p.d.} = \dfrac{\text{energy}}{\text{charge}}$

$V = \dfrac{3.0}{4.0} = 0.75\,\text{V}$

Tip Note that the letter 'V' is used both for the *quantity* p.d. and for the *unit* volts, but the former is italic.

Resistance

All electrical components in a circuit have a resistance. The resistance of a component is defined as:

$$\textbf{resistance} = \frac{\textbf{potential difference}}{\textbf{current}}$$

or

$$R = \frac{V}{I}$$

where R is the resistance of the component, V is the potential difference or voltage across the component and I is the current in the component.

Resistance is measured in **ohms** (Ω). From the definition on p. 19, it follows that:

1 ohm = 1 volt per ampere   or   **1 Ω = 1 V A^{-1}**

The unit of resistance, the ohm, is defined as follows:

The resistance of a component is 1 ohm when the ratio of the potential difference across the component to the current in it is 1 volt per ampere.

You need both p.d. and current to determine the resistance of a component. The circuit diagram below shows how this can be done.

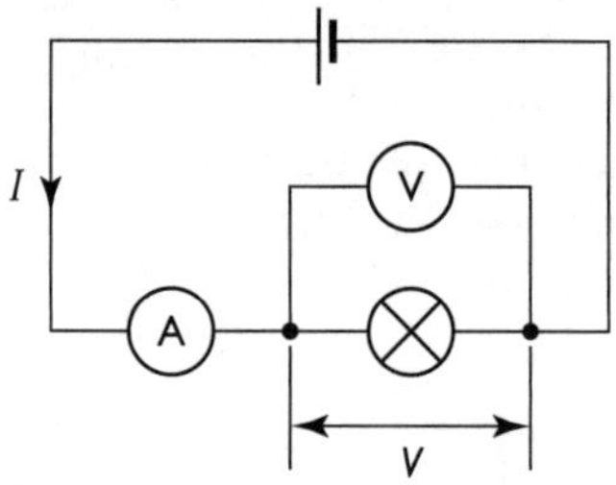

The resistance of the lamp is calculated by dividing the voltmeter reading by the ammeter reading. The ammeter is always connected in series and the voltmeter is always connected in parallel.

Ohm's law

The table below shows the variation of current *I* in a length of metal wire with a potential difference *V* across its ends, when the temperature of the wire is constant. This can be achieved by coiling the insulated wire, placing it in a plastic bag and immersing it in a water bath kept at a constant temperature.

***I*/A**	0	0.20	0.40	0.60	0.80
***V*/V**	0	2.8	5.6	8.4	11.2
***R*/Ω**	–				

The current in the wire is *directly proportional* to the potential difference. The data in the table show that the current doubles from 0.20 A to 0.40 A when the potential difference is doubled from 2.8 V to 5.6 V. The current trebles when the pd is trebled and so on. The ratio of potential difference to current remains constant. We have already named this ratio as 'resistance'. Therefore the resistance of a length of wire kept at constant temperature remains the same. (Using $R = \frac{V}{I}$ shows that the resistance of the above wire is always 14 Ω.) The wire obeys **Ohm's law**, which states:

For a metallic conductor kept at a constant temperature, the current is directly proportional to the potential difference across its ends.

Tip It is a common mistake in examinations to quote the equation $V = IR$ for Ohm's law, rather than giving the precise statement shown above.

I–V characteristics

The *I–V* characteristic of a particular component is a graph of current against potential difference (voltage). For this specification, you must be familiar with the *I–V* characteristics of the following components:

- a resistor kept at constant temperature
- a filament lamp
- a light-emitting diode (LED)

You can use the circuit below to determine the relationship between the current I in a component and the voltage V across it. In the diagram, the component is connected between points **X** and **Y**.

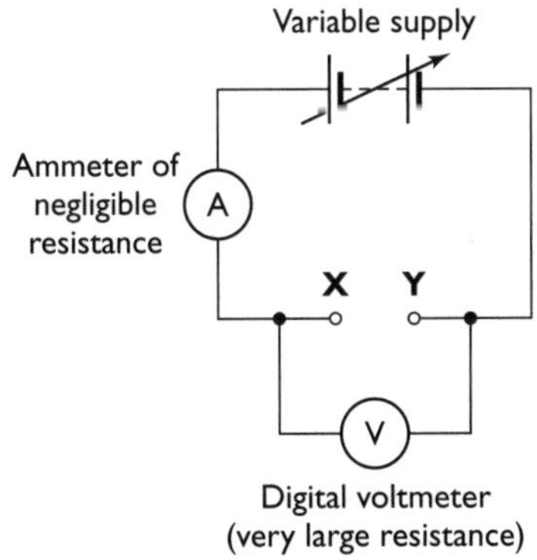

The top three graphs below are the *I–V* graphs for a resistor kept at constant temperature, a filament lamp and a light-emitting diode (LED):

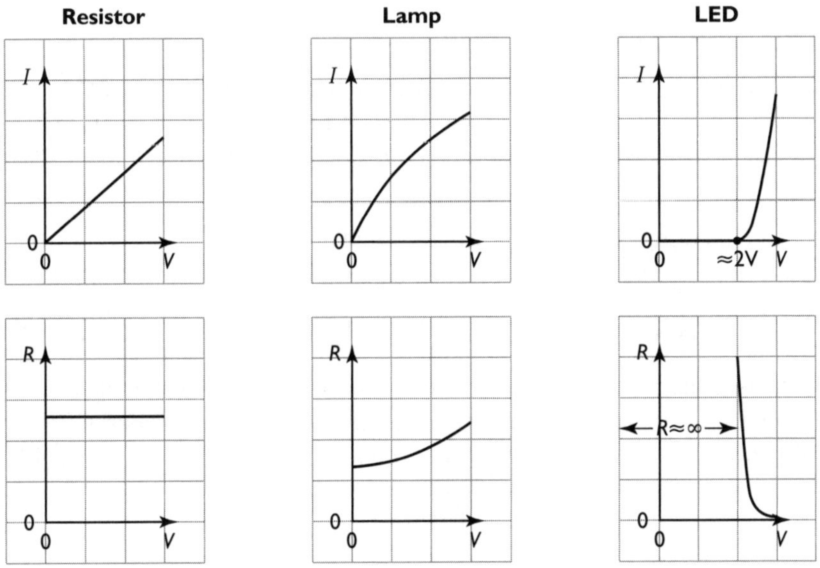

The resistance of a component at any voltage, or current, can be determined using the *I–V* graph and the equation $R = \frac{V}{I}$. In the preceeding set of graphs, the variation of resistance *R* of each component with voltage *V* is shown below each *I–V* graph.

Here are some important facts:

- **Resistor kept at a constant temperature.** The current is directly proportional to the voltage; hence the resistance of the resistor is a constant. A resistor is an **ohmic** component.
- **Filament lamp.** As the voltage across the lamp increases, so does the current. Unlike the resistor, the resistance of the lamp increases as the voltage is increased. Why does this happen? The temperature of the filament increases significantly as the current increases. The metal ions vibrate more quickly. The conduction electrons within the filament have a greater chance of colliding with these ions and hence of losing energy. The mean drift velocity of the electrons decreases and therefore the resistance of the lamp increases. A filament lamp is a **non-ohmic** component.
- **Light-emitting diode (LED).** A semiconductor diode only conducts in one direction. For negative voltages (reverse bias), there is negligible current in the diode and its resistance is infinite. For positive voltages (forward bias), when a certain minimum voltage is exceeded, the LED conducts suddenly and starts to emit light. This minimum voltage is known as the **threshold voltage** of the LED; its value is unique to the colour (wavelength) of light emitted from the LED. It is common practice to use about 2 V for the threshold voltage for an LED. The resistance of the LED decreases dramatically as the voltage across it becomes larger than the threshold voltage. This happens because of an increase in the number density of electrons in the circuit. An LED is a non-ohmic component.

Worked example

The current in a component is 120 mA when the p.d. across it is 2.5 V. When the p.d. across the component is increased to 8.0 V, the current is 300 mA. State and explain whether or not this component is ohmic.

Answer

For an ohmic component, current $\propto$ potential difference

The p.d. increases by a factor of $\frac{8.0}{2.5} = 3.2$

The current increases by a factor of $\frac{300}{120} = 2.5$

Since the current does not increase by the same factor as the potential difference, the current cannot be directly proportional to the potential difference. Therefore, the component is non-ohmic.

Tip There is another way to answer this question and that is to determine the resistance at each potential difference. At 2.5 V the resistance is $\frac{2.5}{0.120} \approx 21\ \Omega$, whereas at 8.0 V the resistance is $\frac{8.0}{0.300} \approx 27\ \Omega$. The component is non-ohmic because the resistance does not remain constant.

Resistivity

Factors that affect resistance

The resistance of a conductor depends on:

- length — long wires have greater resistance than short wires
- cross-sectional area — thin wires have greater resistance than thick wires of the same material
- temperature — the resistance of a wire increases with temperature
- the material of the conductor — for example, a copper wire has less resistance than an aluminium wire of the same dimensions

For a metal wire kept at a constant temperature, the resistance, R, can be determined using the equation:

$$R = \frac{\rho L}{A}$$

where L is the length of the wire, A is the cross-sectional area and ρ (Greek letter rho) is the **resistivity** of the material. The resistivity is a constant for a material (as long as the temperature does not change).

Note: $R \propto L$ and $R \propto \frac{1}{A}$

Resistivity has the unit Ω m. You can show this as follows:

$$\rho = \frac{RA}{L} \rightarrow \left[\frac{\Omega \times \mathrm{m}^2}{\mathrm{m}}\right] \rightarrow \Omega\,\mathrm{m}$$

In examinations, too many candidates confuse the terms 'resistance' and 'resistivity'. Resistance depends on the material, the temperature and its shape; resistivity depends only on the material and temperature.

Worked example

A copper wire in a lawnmower cable has length 15 m and diameter 0.70 mm. Calculate its resistance, given that the resistivity of copper is $1.7 \times 10^{-8}\ \Omega\,\mathrm{m}$.

Answer

$$R = \frac{\rho L}{A}$$

$$R = \frac{1.7 \times 10^{-8} \times 15}{\pi \times (0.35 \times 10^{-3})^2}$$

$R \approx 0.66\,\Omega$

The resistance of the wire is $0.66\,\Omega$.

Tip Do not forget to use the radius of the cable and to convert millimetres into metres.

Role of temperature

Temperature affects the resistivity of a material and hence affects the resistance of a sample.

Pure metals (e.g. copper and iron)

The resistivity of a pure metal *increases* linearly with temperature:

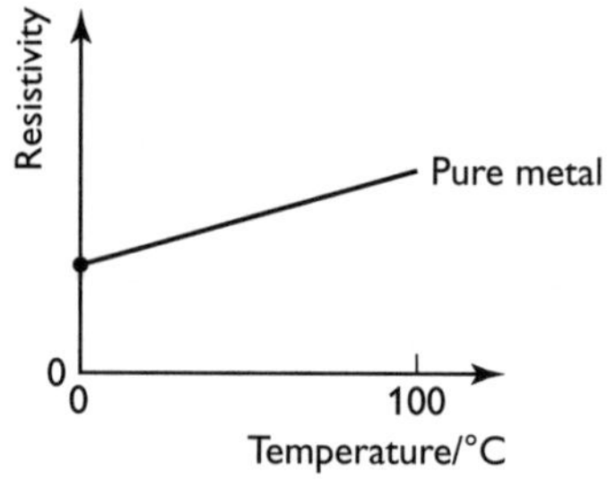

The metal ions vibrate more quickly as the temperature increases and this increases the chance of electrons colliding with these ions. The electrons lose energy and consequently travel with smaller mean drift velocity.

The resistance of a wire made from a pure metal also *increases* as the temperature is increased. The resistance of the wire increases linearly with temperature, so a graph of resistance of a pure metal wire against temperature has the same shape as the resistivity against temperature graph:

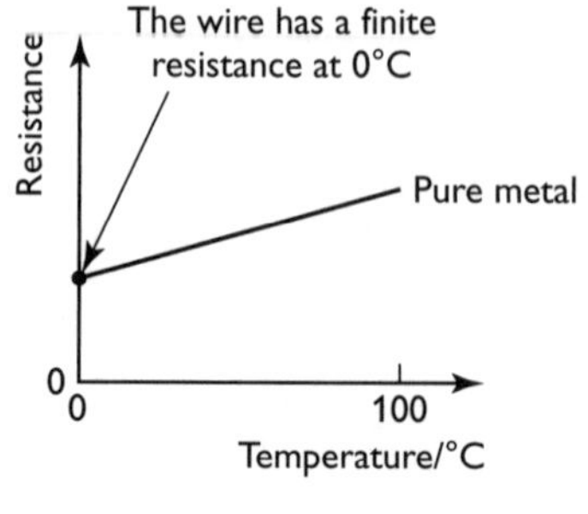

Semiconductors (e.g. silicon)

The resistivity of a semiconductor *decreases* as the temperature increases. As with pure metals, the atoms vibrate more quickly. However, there is a significant increase in the number of free electrons. The number density of the electrons increases with the rise in temperature and this leads to a decrease in the resistivity of the material.

Negative temperature coefficient (NTC) thermistors

The resistance of a thermistor falls as its temperature rises:

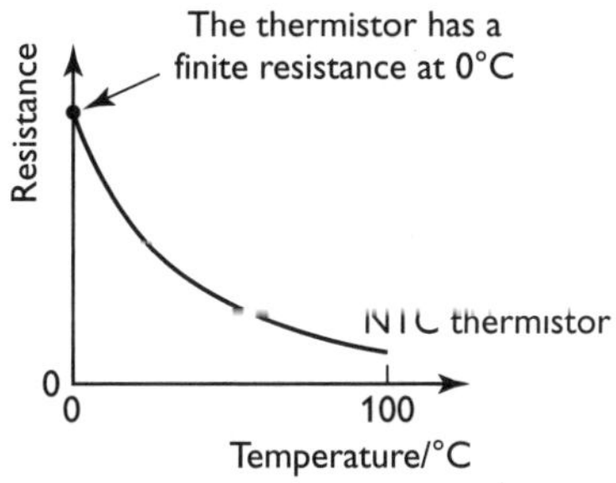

Thermistors are made from semiconductors or metal oxides. An increase in temperature leads to a large increase in the number density of electrons and this lowers the resistance of the component in an electrical circuit.

The table below summarises the properties of the components.

	Pure metal	Semiconductor	NTC thermistor
Rise in temperature *increases* the resistivity and resistance	✓		
Rise in temperature *decreases* the resistivity and resistance		✓	✓

Power

The power equations

A component in a circuit transfers electrical energy to other forms. A resistor dissipates energy as heat whereas a component such as an LED emits light. The rate at which electrical energy is transferred is known as **power**. The word equation for power is:

$$\textbf{power} = \frac{\textbf{energy}}{\textbf{time}}$$

Power is measured in watts (W). From the definition above, it follows that:

1 watt = 1 joule per second or $\mathbf{1\,W = 1\,J\,s^{-1}}$

Tip Do not confuse the *unit*, watt (W), with the *quantity*, energy transfer (*W*), which are both represented by the same letter.

We can derive a useful equation for electrical power by using the definitions of voltage and charge:

energy = voltage × charge (p. 17)

and

charge = current × time (p. 13)

Therefore:

$$\text{power} = \frac{\text{energy}}{\text{time}} = \frac{\text{voltage} \times (\text{current} \times \cancel{\text{time}})}{\cancel{\text{time}}}$$

and thus:

power = voltage × current

The above equation can be written as:

$$\boldsymbol{P = VI}$$

The resistance R of a resistor is given by $R = \frac{V}{I}$. Substituting for V and I into the power equation above gives two useful equations:

$$P = VI = (IR) \times I \qquad \boldsymbol{P = I^2R}$$

and

$$P = VI = V \times \left(\frac{V}{R}\right) \qquad \boldsymbol{P = \frac{V^2}{R}}$$

Worked example

A car headlamp is labelled as '12 V, 24 W'. Determine:

(a) the resistance of the filament

(b) the amount of energy transferred in a time of 2.0 hours

Answer

(a) $V = 12\,\text{V}$; $P = 24\,\text{W}$; $R = ?$

The correct power equation is $P = \frac{V^2}{R}$

Therefore, $24 = \frac{12^2}{R}$

$R = \frac{12^2}{24} = 6.0\,\Omega$

The resistance of the filament is 6.0 Ω.

(b) $\text{power} = \frac{\text{energy}}{\text{time}}$

Therefore, energy = power × time

= 24 × (2.0 × 3600)

energy = $1.728 \times 10^5\,\text{J} \approx 1.7 \times 10^5\,\text{J}$

Tip You can think of power as the 'amount of energy transferred per second'. Hence the energy transferred must be equal to the power multiplied by the number of seconds.

Fuses

A fuse is a safety device used in most domestic appliances. It is constructed from a thin piece of wire that melts quickly when its **current rating** is exceeded. Fuses that fit in three-pin plugs have current ratings of 1 A, 3 A, 5 A and 13 A. The correct fuse for an appliance must have a current rating slightly greater than the normal operating current for the appliance.

Worked example

A 1.8 kW electric kettle is connected to a domestic 230 V mains supply. What is the most appropriate current rating for the fuse?

Answer

$V = 230\,V$; $P = 1800\,W$; $I = ?$

The current in the appliance is I.

$$P = VI$$

$$I = \frac{P}{V}$$

$$= \frac{1800}{230} = 7.8\,A$$

The current rating for the fuse must be greater than 7.8 A. The correct current rating for the fuse is 13 A.

The kilowatt-hour (kW h)

The joule is the SI unit for energy. However, when considering energy transfers for domestic and industrial use, the unit joule is too small. For convenience, most electricity companies use a larger unit, the kilowatt-hour (kW h), for billing.

The kilowatt-hour is defined as follows:

One kilowatt-hour is the energy transferred by a 1 kW device operated for a time of 1 hour.

What is 1 kilowatt-hour in joules?

energy = power × time

$1\,kW\,h = 1000\,W \times 3600\,s$

$\mathbf{1\,kW\,h = 3.6 \times 10^6\,J}$

The amount of energy in kW h is determined as follows:

number of kW h = power in kW × time in hours

Hence the cost of operating a domestic appliance can be found using the following relationship:

cost = (number of kW h) × (cost per kW h)

Worked example

A student watches her television for an average of 3.0 hours each day. The television has a power rating of 55 W. The cost of each kW h is 8.2 p. Calculate the cost towards the electricity bill over a period of one year.

Answer

number of kW h = power (kW) × time (h)

number of kW h = 0.055 × (3.0 × 365) = 60.225

cost = (number of kilowatt-hours) × (cost per kW h)

cost = 60.225 × 8.2 ≈ 494 p

The cost of watching the television over the year is £4.94.

Tip The cost comes out the same if you use 1 year = $365\frac{1}{4}$ days.

Series and parallel circuits

Kirchhoff's second law

You have already met Kirchhoff's first law (p.14). Kirchhoff's second law states:

The sum of the voltages (p.d.) round a closed loop in a circuit is equal to the sum of electromotive forces (e.m.f.) in that loop.

This law is an expression of the conservation of *energy*. A mathematical version for Kirchhoff's second law is:

Σe.m.f. = Σp.d. *or* Σe.m.f. = $\Sigma(IR)$

Note: The Greek letter sigma, Σ, is shorthand for 'sum of'.

Worked example

The diagram shows an electrical circuit consisting of resistors and cells. The values of the electromotive force for the cells are 1.50 V and 0.80 V. Calculate the current in the resistor of resistance 120 Ω.

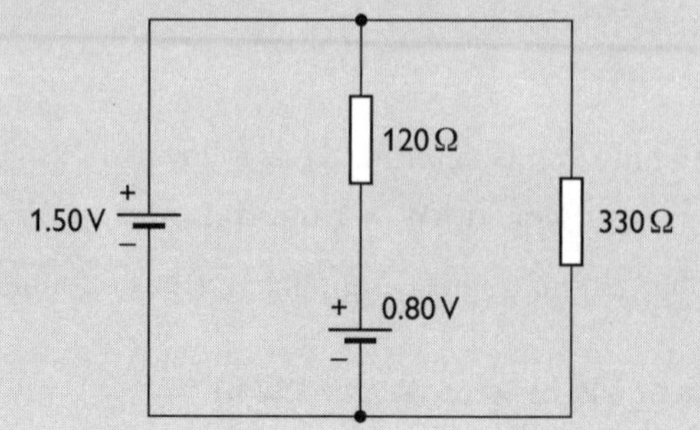

Answer

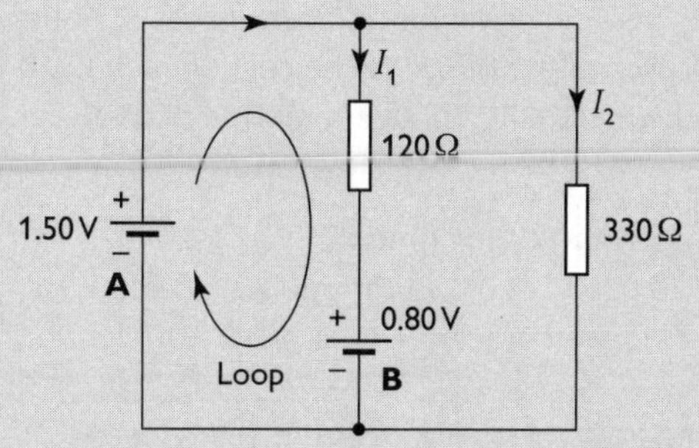

Cell **A** has the larger electromotive force, hence the current I_1 in the 120 Ω resistor is in the direction shown in the above diagram. Kirchhoff's second law is applied to the circuit 'loop', which consists of the two cells and the 120 Ω resistor.

According to Kirchhoff's second law, we have:

$$\Sigma \text{e.m.f.} = \Sigma \text{p.d.}$$

$$1.50 - 0.80 = I_1 \times 120$$

$$I_1 = \frac{1.50 - 0.80}{120}$$

$$I_1 \approx 5.8 \times 10^{-3}\,\text{A}\ (5.8\,\text{mA})$$

Tip The electromotive force values have to be subtracted because of the 'opposing' polarities of the cells in the clockwise loop shown.

Rules for series and parallel circuits

The diagram below shows three resistors of resistance R_1, R_2 and R_3 connected in series to a supply:

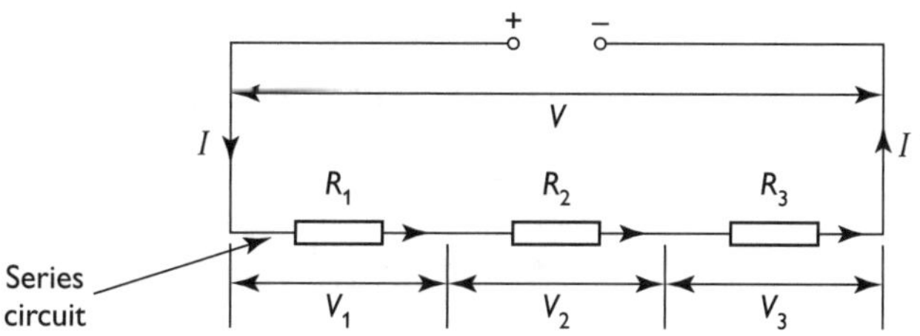

Here are some rules for a **series circuit**:

- We can apply the equation $V = IR$ to each resistor or to the whole circuit.
- The current, I, is the same in each resistor.
- The total potential difference, V, across the resistors is equal to the sum of the potential differences across the resistors, i.e. $V = V_1 + V_2 + V_3 \ldots$
- The total resistance, R, of the circuit is given by the equation $\boldsymbol{R = R_1 + R_2 + R_3 \ldots}$

The diagram shows three resistors of resistances R_1, R_2 and R_3 connected in parallel to a supply:

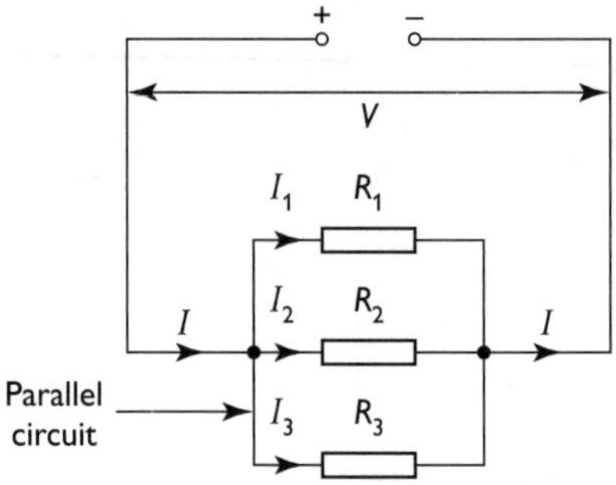

Here are some rules for a **parallel circuit:**

- We can apply the equation $V = IR$ to each resistor or to the whole circuit.
- The potential difference, V, across each resistor is the same.
- The total current, I, is the sum of the currents in each resistor; i.e. $I = I_1 + I_2 + I_3 \ldots$. (This is in accordance with Kirchhoff's first law.)
- The total resistance, R, of the circuit is given by the equation:

$$\frac{1}{R} = \frac{1}{R_1} + \frac{1}{R_2} + \frac{1}{R_3} + \ldots$$

In exams it is safer to use this reciprocal equation for the total resistance. With *two* resistors, a simpler version for the total resistance is the equation:

$$R = \frac{R_1 R_2}{R_1 + R_2}$$

Note that this equation can be used *only* for two resistors. It cannot be extended easily to three or more resistors.

Worked example

For the circuit below, calculate the current I drawn from the battery. The electromotive force of the battery is 6.0 V and it has negligible internal resistance (p. 31).

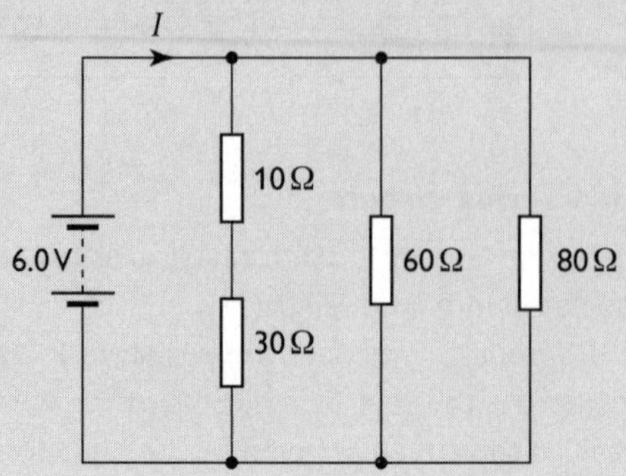

Answer

We need to determine the total resistance of the two resistors in series before we can use the reciprocal equation.

total 'series' resistance $= R_1 + R_2 = 10 + 30 = 40\,\Omega$

The total resistance of the circuit is R.

$$\frac{1}{R} = \frac{1}{R_1} + \frac{1}{R_2} + \frac{1}{R_3}$$

$$\frac{1}{R} = \frac{1}{40} + \frac{1}{60} + \frac{1}{80} = 0.0541666...$$

$$R = \frac{1}{0.054166...} = 18.46\,\Omega \approx 18\,\Omega$$

The circuit current I can be determined using $V = IR$, where R is the total resistance of the circuit:

$V = IR$; $6.0 = I \times 18.46$

$$I = \frac{6.0}{18.46} \approx 0.33\text{ A}$$

The current I is 0.33 A.

Internal resistance

All sources of electromotive force — for example, chemical cells, solar cells and dynamos — have an internal resistance. In the case of a chemical cell, this resistance is due to the chemical substance through which the charges have to travel. Charges travelling through the cell release energy as heat and the cell will warm up. This implies that a cell can never be 100% efficient at transferring all its energy to the components connected across its terminals.

The diagram below shows how we can model a chemical cell (or a power supply). The internal resistance is represented by an internal resistor of resistance r that is connected in *series* with cell of electromotive force E.

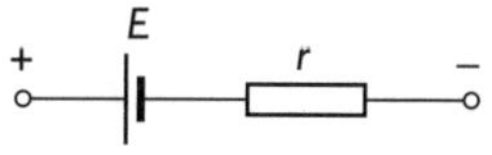

The circuit diagram below shows a cell (or power supply) connected to an external resistor of resistance R.

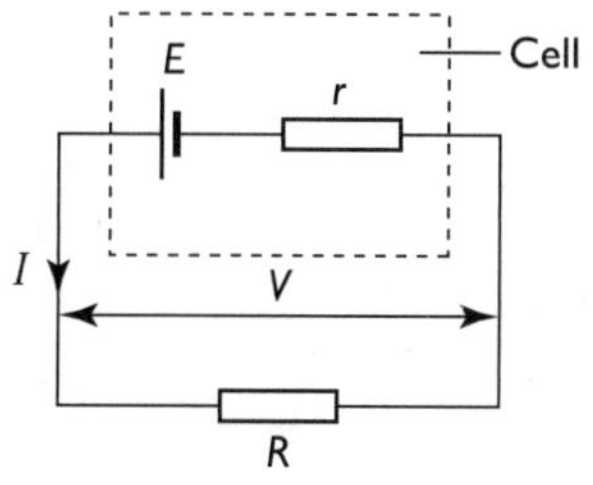

The current I travels through both the internal and the external resistors; hence there is a potential difference across each resistor. The potential difference V across the external resistor is the same as the potential difference across the terminals — this is the **terminal potential difference**.

According to Kirchhoff's second law, we have:

$$\Sigma \text{e.m.f.} = \Sigma \text{p.d.}$$

$$E = IR + Ir$$

$$E = V + Ir$$

$$\boldsymbol{V = E - Ir}$$

The term 'Ir' represents the potential across the internal resistance or the '**lost volts**'. For any source of electromotive force providing a current to an external circuit, the terminal potential difference V is *less* than the electromotive force E, unless the current is zero or negligible, in which case $V \approx E$. This is illustrated in the worked example below.

Worked example

A variable resistor of maximum resistance 100 Ω is connected across the terminals of a cell. The electromotive force of the cell is 1.50 V and its internal resistance is 0.30 Ω. With the help of a table, deduce what happens to the current in the circuit and to the terminal potential difference across the cell as the resistance of the variable resistor is altered from zero to 100 Ω.

Answer

The current I can be determined as follows:

$$E = IR + Ir = I(R + r)$$

$$I = \frac{E}{R+r}$$

The terminal p.d. V is given by the equation:

$$V = E - Ir$$

The table below shows the variation of I and V with external resistance R.

As the external resistance increases, the current in the circuit decreases and the terminal potential difference gets closer to the electromotive force of the cell.

R/Ω	$I = \frac{1.50}{R+0.30}$/A	$V = 1.50 - (I \times 0.30)$/V
0	5.00	0
0.15	3.33	0.50
0.30	2.50	0.75
1.00	1.15	1.15
20.0	0.074	1.48
100.0	0.015	≈ 1.50

Tip The maximum current from the cell is 5.00 A. This occurs when the terminals are 'shorted out'; hence $R = 0$. When $R = 0$, $V = 0$ and therefore the potential difference across the internal resistor is 1.50 V. All the energy from the cell is wasted as heat in the cell itself.

Power dissipated in an external resistor

A cell delivers power to any component that is connected across its terminals. There is *maximum* power transferred from the cell to the external component when the resistance of the component is *equal* to the internal resistance of the cell. Consider a cell of electromotive force (E) 1.5 V and internal resistance (r) 0.8 Ω, connected across a variable resistor. The resistance of the variable resistor is R.

The current I in the variable resistor is given by the equation:

$$I = \frac{E}{R+r}$$

The power P dissipated in the variable resistor is given by equation:

$$P = I^2R$$

The table below shows what happens to the power dissipated in the variable resistor as its resistance is increased in the range 0 to 10 Ω.

R/Ω	$I = \frac{1.5}{R+0.8}$/A	P/W
0	1.875	0
0.2	1.500	0.450
0.4	1.250	0.625
0.8	0.938	0.703
1.0	0.833	0.694
1.4	0.682	0.651
10	0.139	0.193

Note that the power dissipated is maximum when $R = r$.

The graph below shows how the power P that is dissipated in the variable resistor changes with its resistance R.

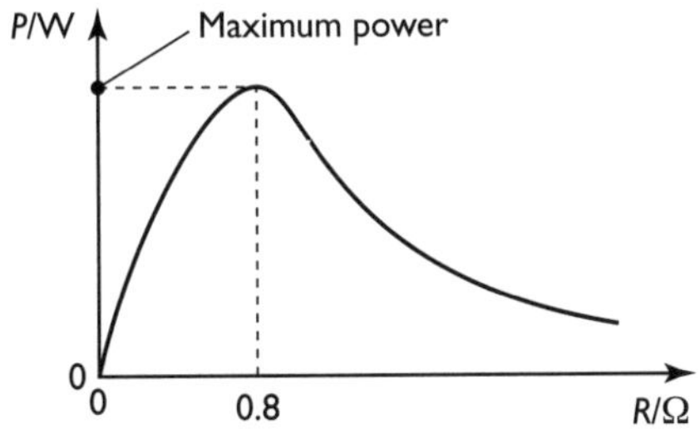

Determining internal resistance in the laboratory

The circuit below can be used to determine the internal resistance r of a cell (or a power supply).

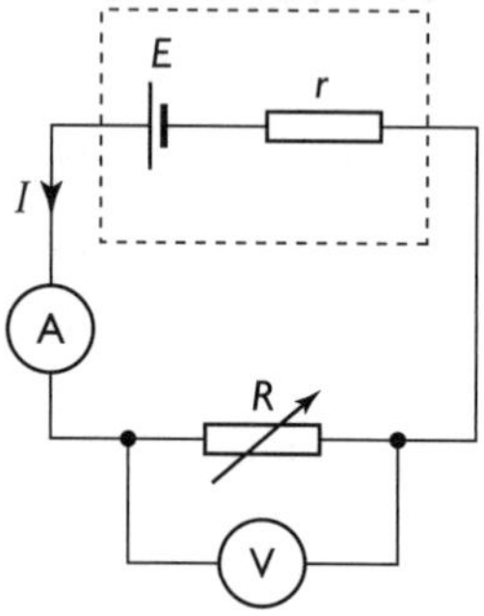

The terminal potential difference V across the cell is measured using a high-resistance digital voltmeter. The current I is measured using a low-resistance ammeter placed in series with a variable resistor. The resistance of the variable resistor is altered so as to obtain a range of values of V and I.

The internal resistance and electromotive force of the cell can be deduced from a graph of the terminal potential difference V against current I.

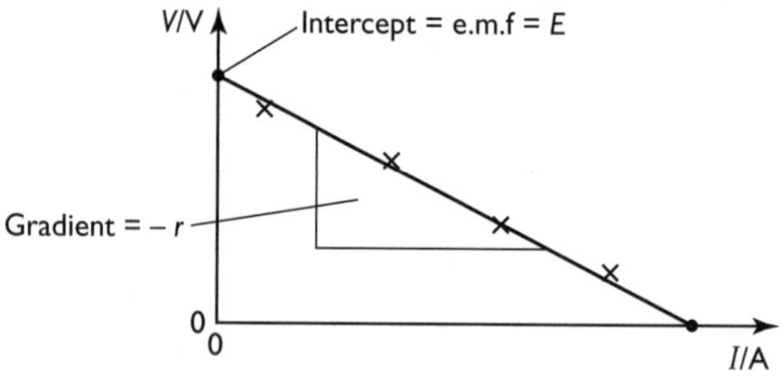

The terminal potential difference is given by the equation (p. 32):

$$V = E - Ir$$

When this equation is compared with the equation for a straight line ($y = mx + c$) we have:

- **gradient = $-r$**
- **intercept with V-axis = E**

Practical circuits

Light-dependent resistors

The resistance of a light-dependent resistor (LDR) falls as the intensity of the incident light increases:

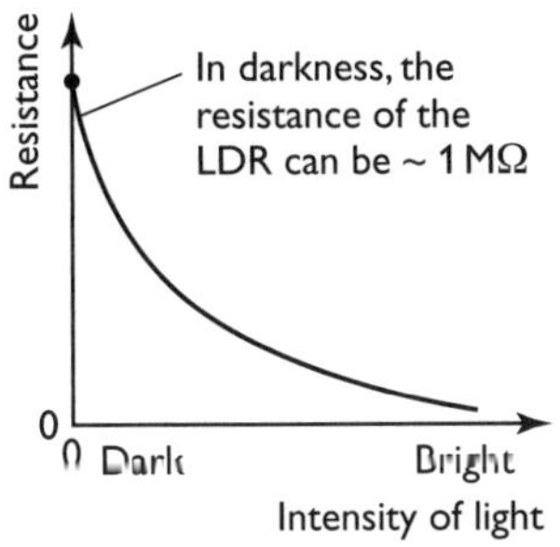

LDRs are made using semiconductors. An increase in the intensity of light frees an increased number of electrons from the atoms. This leads to a huge increase in the number density of electrons and lowers the resistance of the component in an electrical circuit.

Potential divider circuits

In its simplest form, a potential divider circuit consists of two components (e.g. resistors, LDRs, thermistors, LEDs) connected in series to a supply, with the 'output voltage' taken across one of the components. A potential divider circuit with two resistors of resistances R_1 and R_2 is shown below.

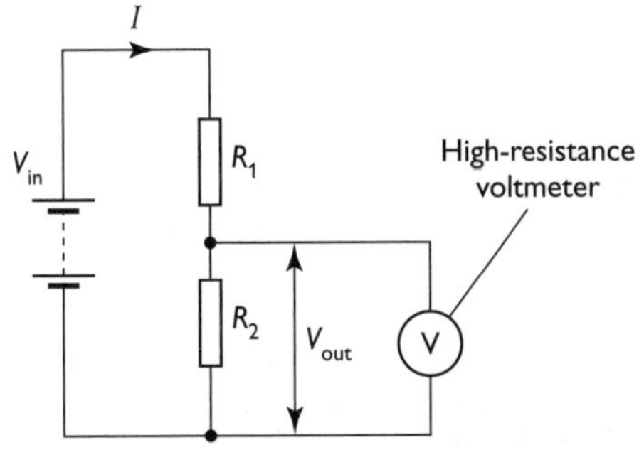

The supply has negligible internal resistance. The electromotive force of the supply is V_{in}. The voltmeter connected across the resistor of resistance R_2 has an infinite resistance. The voltmeter can be a digital voltmeter that draws negligible current from the potential divider circuit.

The circuit current I is given by the equation:

$$I = \frac{V_{in}}{R_1 + R_2}$$

The current in each resistor is the same. The output voltage V_{out} across the resistor of resistance R_2 is given by:

$$V_{out} = IR_2$$

$$V_{out} = \frac{V_{in}}{R_1 + R_2} \times R_2$$

This simplifies to the equation below, which is called the **potential divider equation**:

$$\boldsymbol{V_{out} = \left(\frac{R_2}{R_1 + R_2}\right) \times V_{in}}$$

Since the current in each component is the same, this implies that the potential difference across a component is directly proportional to its resistance:

$$V = IR \propto R$$

Therefore:

$$\boldsymbol{\frac{V_1}{V_2} = \frac{R_1}{R_2}}$$

where V_1 is the voltage across the resistor of resistance R_1 and V_2 is the voltage across the resistor of resistance R_2 (see the diagram below). Note that this equation is a variation of the original potential divider equation.

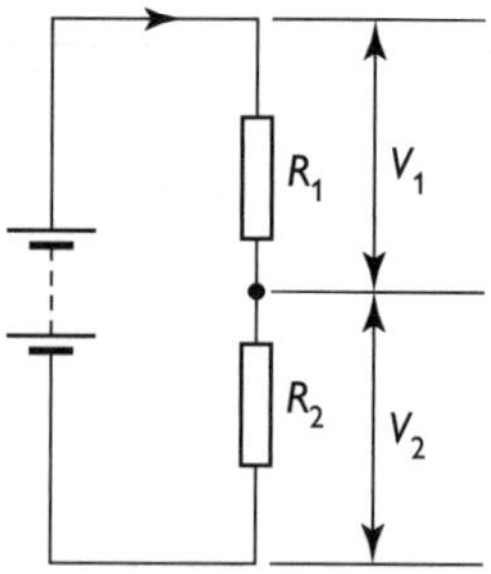

Worked example

The diagram below shows a circuit designed by a student in order to monitor the light intensity in a room.

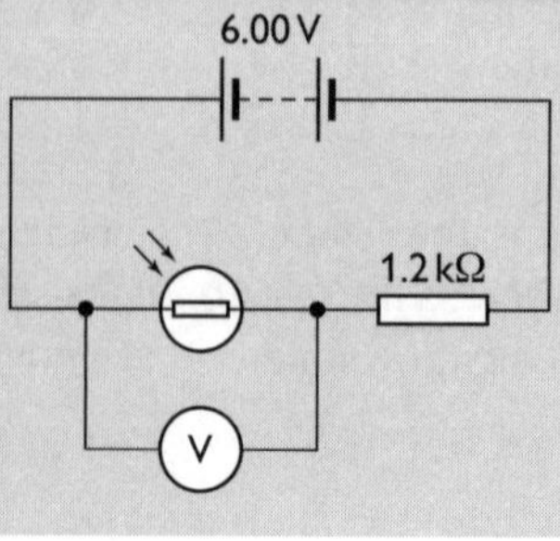

The electromotive force of the battery is 6.00 V and its internal resistance is negligible. The voltmeter placed across the LDR has an infinite resistance. In direct sunlight, the resistance of the LDR is 510 Ω.

(a) Determine the voltmeter reading in direct sunlight.

(b) Describe what will happen to the voltmeter reading as the intensity of light decreases.

Answer

(a) $R_1 = 1.2\,\text{k}\Omega$; $R_2 = 510\,\Omega$; $V_{in} = 6.00\,\text{V}$

$$V_{out} = \left(\frac{R_2}{R_1 + R_2}\right) \times 6.00$$

$$V_{out} = \left(\frac{510}{510 + 1200}\right) \times 6.00$$

$$V_{out} = 1.79\,\text{V}$$

The voltmeter reading in direct sunlight is 1.79 V.

(b) The resistance of the LDR increases as the intensity of light decreases. Therefore, the voltmeter reading will increase because the potential difference across the LDR is directly proportional to its resistance.

Tip The voltage across the LDR increases because it takes a greater share of the input voltage to the potential divider circuit.

Wave motion

Progressive waves

A **wave** is a disturbance in the form of vibrations travelling through either space or a substance. Electromagnetic waves (p. 45) can travel through a vacuum, but many waves require some material (medium) for their propagation. Waves that require a medium for their transmission are known as **mechanical waves**, for example:

- sound waves
- water waves
- waves in a stretched string or rope
- seismic waves

A **progressive wave** transfers energy from one place to another. When a wave travels through a material, it makes the particles of that material vibrate. Wave motion is the result of these vibrating particles affecting the neighbouring particles. It is important to appreciate that the particles themselves do not travel in the direction of energy transfer.

Longitudinal and transverse waves

A **longitudinal wave** is a wave in which the particles of the medium vibrate or oscillate *parallel* to the direction of the wave velocity.

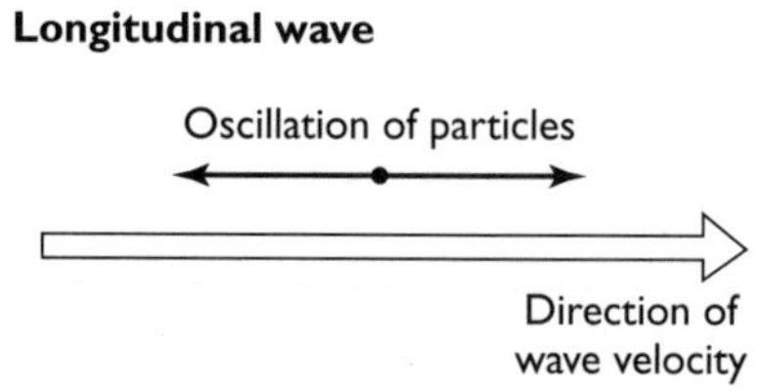

A slinky may be used to demonstrate a longitudinal wave:

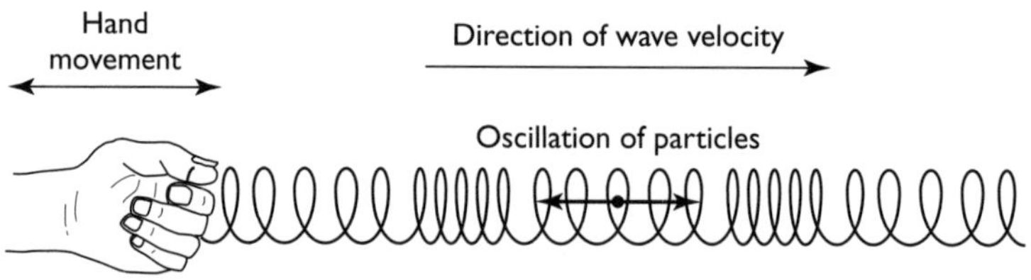

Sound is a longitudinal wave created by any object that is vibrating in air. It is transmitted through a series of **compressions** (C) and **rarefactions** (R):

- Compressions are regions of higher than normal air pressure where the air particles have been squeezed together.
- Rarefactions are regions of lower than normal air pressure where the air particles have been moved further apart.

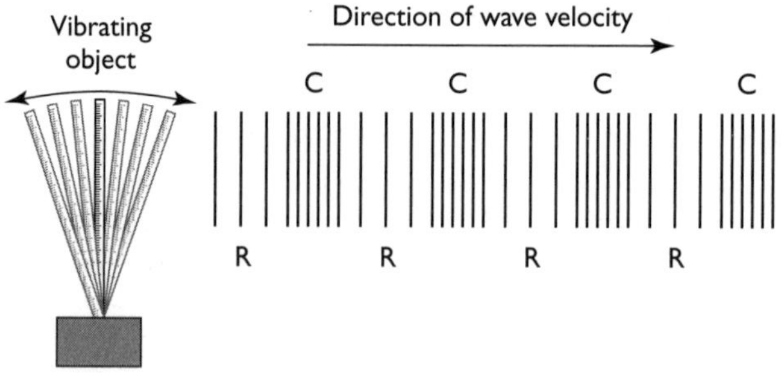

The series of compressions and rarefactions travels through still air at a speed of about $340\,m\,s^{-1}$.

A **transverse wave** is a wave in which the particles of the medium vibrate or oscillate at right angles to the direction of the wave velocity, as shown below.

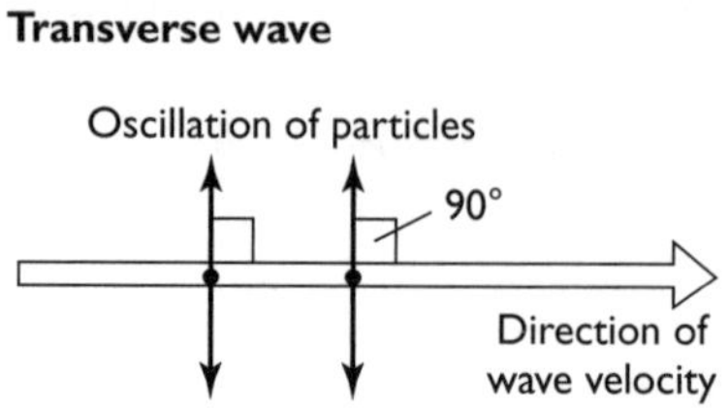

A slinky may also be used to demonstrate a transverse wave:

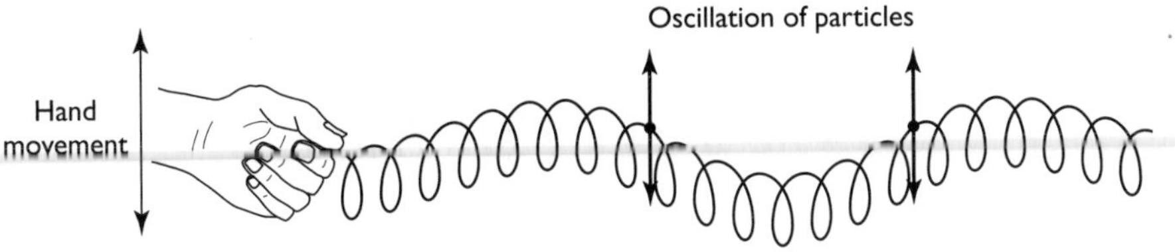

Surface water waves, electromagnetic waves (e.g. light) and waves on a stretched string or rope are good examples of transverse waves.

Wave quantities

It is important to learn the following definitions of the terms that describe wave motion:

- The **displacement** of a wave is the distance and direction of a vibrating particle from its equilibrium position. Displacement is measured in metres (m).
- The **amplitude** of a wave is the maximum displacement of a vibrating particle of the medium. For a transverse wave, the amplitude is the distance of the **peak** of the wave (crest) or the **trough** of the wave from the equilibrium position.
- The **wavelength** of a wave is the distance between two adjacent points oscillating in phase (in step). For a transverse wave, the wavelength of the wave is the distance between adjacent peaks or troughs. For a longitudinal wave, the wavelength is the distance between adjacent compressions or rarefactions. Wavelength is measured in metres (m). The displacement against distance graph below shows how the amplitude A of the wave and its wavelength λ can be determined.

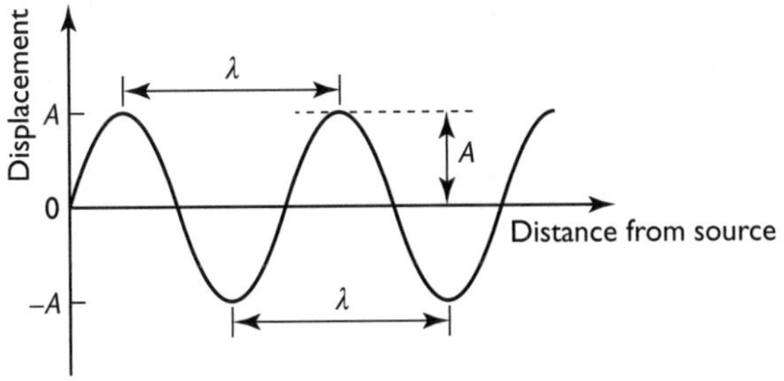

- The **period** of a wave is the time taken for a particle of the medium to execute one complete oscillation. It is also the time taken for the wave to progress one whole wavelength along the direction of the wave velocity. Period is measured in seconds (s). The displacement against time graph below shows how the amplitude A of the wave and its period T can be determined.

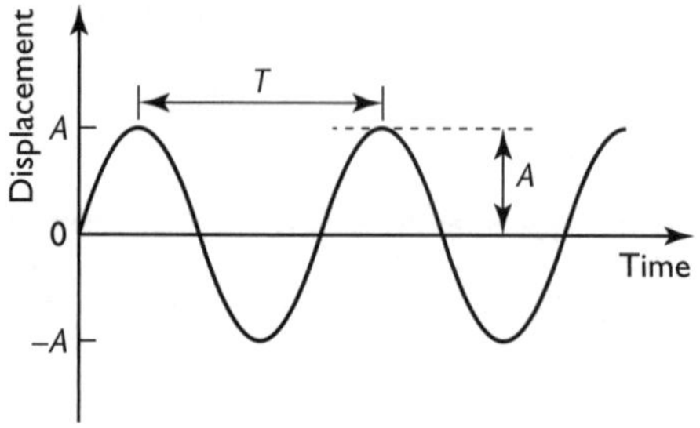

- The **frequency** of a wave is the number of oscillations of a particle of the medium per unit time. The unit of frequency is the hertz (Hz):

  1 Hz = 1 oscillation per second

Period and frequency are related by the following equation:

$$\textbf{frequency} = \frac{\textbf{1}}{\textbf{period}}$$

$$f = \frac{1}{T}$$

Phase, anti-phase and phase difference

Two points on a wave that have the same displacement and direction of oscillation are said to be **in phase**. Points on a wave that are separated by a distance of one wavelength (λ), or by a whole-number multiple of the wavelength, are oscillating in phase.

Two points on a wave are in **anti-phase** when they have the same magnitude of displacement but the direction of oscillation is opposite. Points separated by a distance of half a wavelength, or an odd number of half wavelengths, are oscillating in anti-phase.

The **phase difference** between two oscillating particles on a wave is the fraction of an oscillation between the oscillations of the two particles. Phase difference is measured in degrees or radians.

1 oscillation → 360° → 2π radians

Phase difference depends on the separation between the two points on a wave. This is illustrated in the table below.

Separation between points	Phase difference
$\frac{\lambda}{4}$	90°
$\frac{\lambda}{2}$	180°
$\frac{3\lambda}{2}$	270°
x	$\frac{x}{\lambda} \times 360°$

The general equation for determining phase difference is:

$$\textbf{phase difference} = \frac{\textbf{separation between points} \times 360°}{\textbf{wavelength}}$$

Worked example

A wave has a wavelength of 12 cm. Calculate the phase difference between two points on the wave that are separated by 4.0 cm.

Answers

$$\text{phase difference} = \frac{x}{\lambda} \times 360°$$

$$\text{phase difference} = \frac{4.0}{12} \times 360° = 120°$$

The phase difference between the two points with a separation of 4.0 cm is 120°.

Wave speed

The speed v of a wave can be determined from its wavelength λ and frequency f. In a time equal to one period, T, the entire wave travels a distance equal to one wavelength λ in the direction of the wave velocity. The speed v of the wave is given by:

$$v = \frac{\text{distance}}{\text{time}} = \frac{\lambda}{T}$$

However, frequency $f = \frac{1}{T}$ so $\boldsymbol{v = f\lambda}$

The equation $\boldsymbol{v = f\lambda}$ is the **wave equation**. It can be applied to all periodic waves.

Worked example

The audible range for a healthy human ear is about 20 Hz to 20 kHz. The speed of sound is $340\,\text{m}\,\text{s}^{-1}$ in air. Calculate the maximum and minimum wavelengths of sound that can this ear can register.

Answer

The wave equation is $v = f\lambda$.

The wavelength is therefore given by $\lambda = \frac{v}{f}$.

The wavelength is inversely proportional to the frequency.

The maximum wavelength λ_{max} corresponds to the lowest frequency. Hence:

$$\lambda_{max} = \frac{340}{20} = 17\,\text{m}$$

The minimum wavelength corresponds to the highest frequency. Hence:

$$\lambda_{min} = \frac{340}{20 \times 10^3} = 0.017\,\text{m}\ (1.7\,\text{cm})$$

Tip Do not forget that the frequency is given in kHz and has to be converted into Hz.

Some wave properties

All waves can be reflected, refracted and diffracted and all show interference (p. 48).

Reflection is the 'bouncing' back of a wave at a surface. When a wave (e.g. of sound or light) is incident on a plane surface, the angle of incidence is equal to the angle of reflection. Note that the angles are measured relative to the normal.

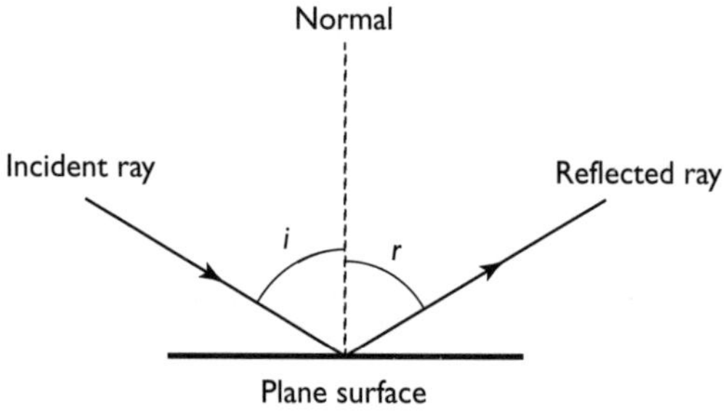

i = angle of incidence
r = angle of reflection

Refraction is the bending of a wave at the interface of two media. Refraction occurs because there is a change in the speed of a wave. The frequency of the wave remains the same as it travels from one medium into another medium.

The speed of light *decreases* as it travels from air into glass; the ray of light is refracted *towards* the normal.

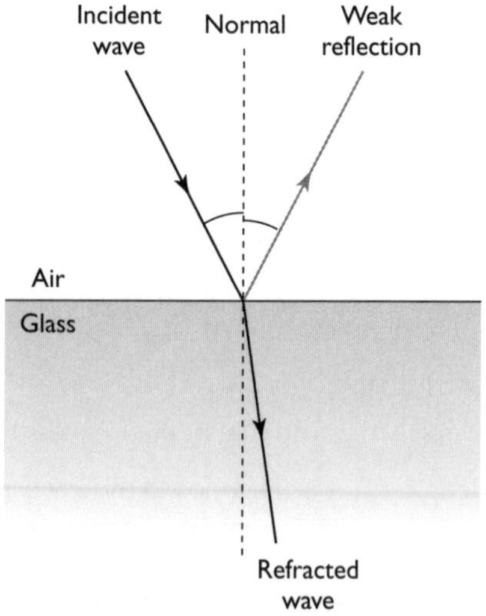

The speed of surface water waves depends on the depth of water. These waves travel more slowly in shallow water than in deep water. Water surface waves are refracted *away* from the normal as they travel from shallow water into deeper water.

Note: The frequency of the water waves remains the same in deep water and in shallow water.

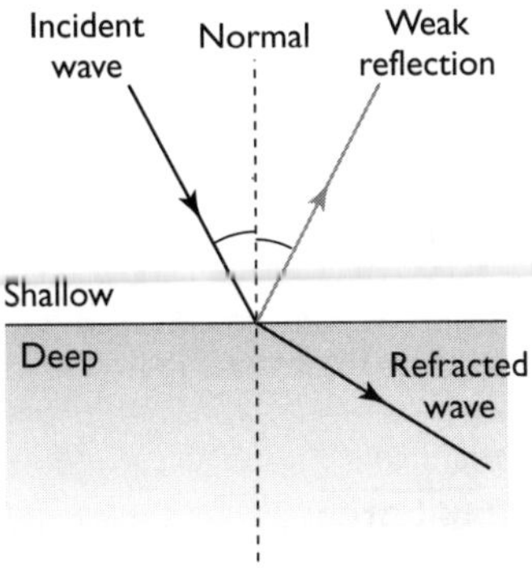

Diffraction is the spreading of a wave at a slit or round an object. The amount of spreading of a wave at a slit depends on the relative width of the slit and the wavelength of the incident wave. The diagrams below show what happens to parallel wavefronts of a wave encountering slits of different widths. Note that a **wavefront** is a line joining all the points of the wave that are in phase. The separation between adjacent wavefronts is equal to the wavelength of the wave.

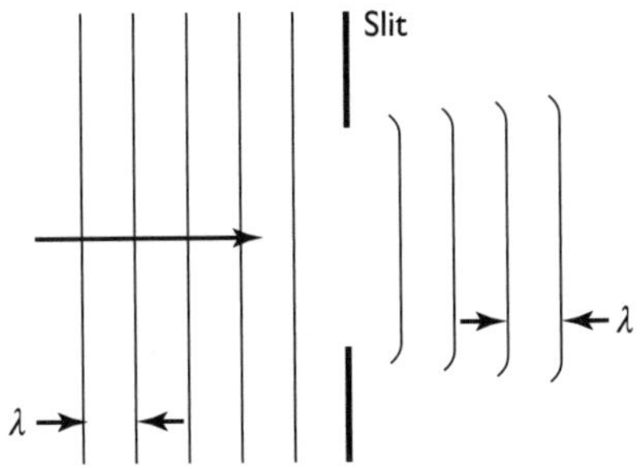

Width of slit >> wavelength

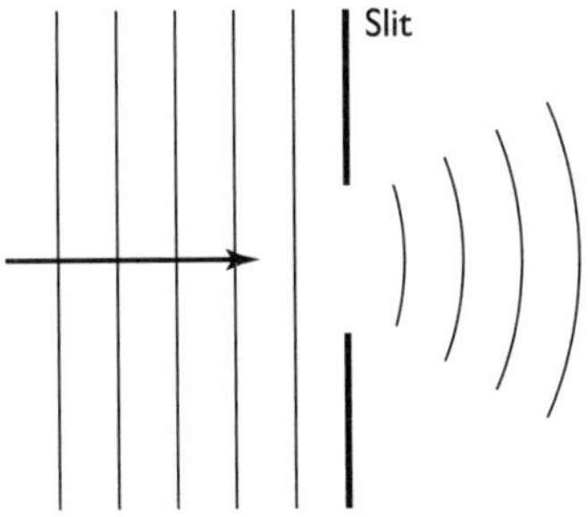

Width of slit > wavelength

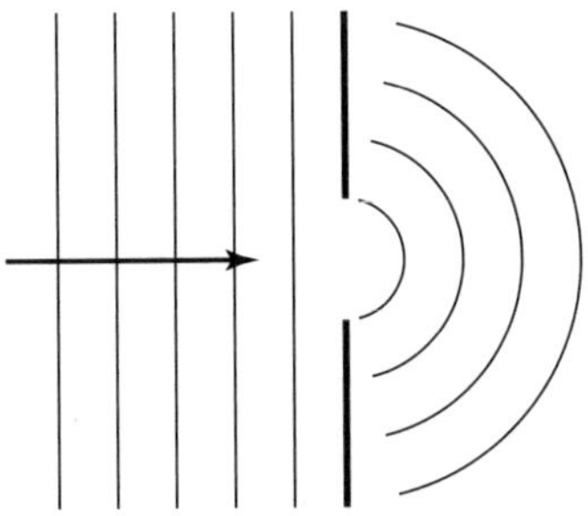

Width of slit ≈ wavelength

Important points about diffraction include:

- Whatever the size of the slit, there is no change in the wavelength in the wave as it emerges from the slit.
- There is hardly any diffraction when the width of the slit is much greater than the wavelength.

- Diffraction effects are more prominent when the width of the slit is comparable to the wavelength. When the width of the slit is roughly equal to the wavelength, the slit behaves like a **point source** spreading circular wavefronts from the slit.

Intensity and amplitude

The **intensity** of a wave at a point in space has a precise meaning. It is defined as the radiant power transmitted per unit cross-sectional area at right angles to the velocity of the wave:

$$\textbf{intensity} = \frac{\textbf{power}}{\textbf{cross-sectional area}}$$

Intensity is measured in watts per square metres ($W\,m^{-2}$).

As a wave spreads out from a source (e.g. light from a star or lamp, circular ripples on the surface of water when a stone is dropped), its amplitude decreases. The intensity I of a wave and the amplitude A of the wave are related as follows:

$$I \propto A^2$$

The intensity is directly proportional to the square of the amplitude. This means that, for example, if the amplitude of a wave decreases by a factor of 10, the intensity of the wave decreases by a factor of 10^2, i.e. 100.

Worked example

(a) Waves from a source have an amplitude of 2.0 mm and an intensity of $60\,W\,m^{-2}$. The amplitude of the waves decreases as they spread out from the source. What is the intensity of the waves where the amplitude is 0.1 mm?

(b) Lasers produce intense light. The radiant power of a certain laser is 0.12 mW and it produces a beam of diameter 5.0 mm. Calculate the intensity of the laser light.

Answer

(a) Since $I \propto A^2$ we have:

$$\frac{\text{intensity}}{\text{amplitude}^2} = \text{constant}$$

$$\frac{I}{0.1^2} = \frac{60}{2.0^2} \quad \text{so } I = \frac{60 \times 0.1^2}{2.0^2} = 0.15\,W\,m^{-2}$$

Tip There is no need to convert the amplitudes into metres because we are dealing with ratios and both amplitudes are given in millimetres. Since the amplitude decreases by a factor of $\frac{2.0}{0.1} = 20$, the intensity must decrease by a factor of 20^2:

$$\text{intensity} = \frac{60}{20^2} = 0.15\,W\,m^{-2}$$

(b) The intensity of light from the laser can be calculated using the equation

$$\text{intensity} = \frac{\text{power}}{\text{cross-sectional area}}$$

Therefore:

$$I = \frac{P}{\pi r^2} = \frac{0.12\times10^{-3}}{\pi\times(2.5\times10^{-3})^2} \approx 6.1\,\mathrm{W\,m^{-2}}$$

Tip In order to calculate the intensity in $\mathrm{W\,m^{-2}}$, you must have the power in watts (W) and the radius of the beam in metres (m).

Electromagnetic waves

Light is an electromagnetic wave. All electromagnetic waves:

- can travel through a vacuum
- travel at the speed of speed of light, c, which is $3.0 \times 10^8\,\mathrm{m\,s^{-1}}$
- are transverse waves
- travel through space as oscillating electric and magnetic fields
- can be reflected, refracted, diffracted, polarised and show interference effects

The diagram below shows the electromagnetic spectrum and the typical wavelengths of the different regions.

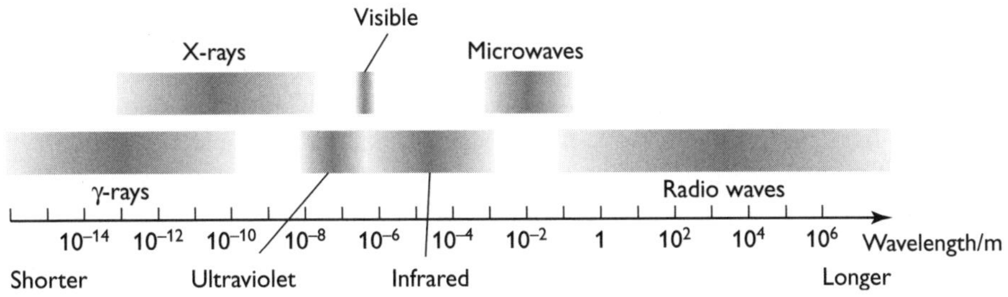

The *only* difference between the different regions of the electromagnetic spectrum is the wavelength, and hence frequency, of the waves. The wavelength λ is related to the frequency f by the wave equation $c = f\lambda$.

Typical wavelengths and practical uses of electromagnetic waves

Electromagnetic waves	Typical wavelength	Use
Gamma rays	10^{-15} m	Medical imaging; sterilisation
X-rays	10^{-10} m	Medical imaging
Ultraviolet	10^{-8} m	Sterilisation; sun-tanning
Visible	4×10^{-7} m (violet) to 7×10^{-7} m (red)	Photography
Infrared	10^{-6} m	Remote controls for DVDs; night-vision camera
Microwaves	1 cm	Mobile phones; microwave cookers
Radio waves	1 km	Radio and television transmission; radio astronomy

Ultraviolet radiation

Ultraviolet radiation is both harmful and beneficial — it depends on the wavelength. The dangers of exposure to this radiation can be minimised by putting on sunscreen. This contains titanium dioxide to absorb the ultraviolet radiation. The three types of ultraviolet radiation are summarised in the table below.

Ultraviolet radiation	Wavelength/10^{-7} m	Characteristic
UV-A	3.2–4.0	Causes wrinkling of the skin
UV-B	2.8–3.2	Damages DNA in skin cells and can trigger cancer; produces vitamin D in skin
UV-C	~0.1–2.8	Almost all is absorbed by the ozone layer

Polarisation

All electromagnetic waves are transverse waves. The electric and magnetic fields oscillate at right angles to the direction in which the wave travels.

In normal light, the electric field oscillates in all directions in a plane that is perpendicular to the direction of travel. Such light is **unpolarised**. When unpolarised light passes through a polarising filter (Polaroid), the transmitted light is **plane polarised** — it has an electric field that oscillates in just one plane. Polaroid consists of long chains of molecules that absorb energy from the oscillating electric field. Light with an electric field oscillating at 90° to the molecular chains is unaffected and hence is transmitted through the Polaroid.

A second Polaroid (or **analyser**) is held beyond the first Polaroid (the **polariser**) and rotated until the transmitted light has maximum intensity. Under these conditions, the long molecular chains of both Polaroids are aligned in the same direction. When the analyser is rotated, the intensity of the transmitted light decreases. When the analyser is rotated through 90°, the intensity of the transmitted light is virtually zero. This is illustrated in the diagram below.

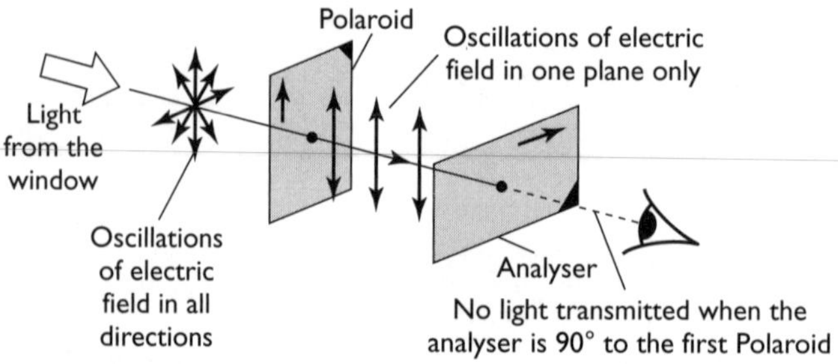

Polarisation of microwaves and radio waves

Only *transverse* waves can be plane polarised. This implies that all electromagnetic waves can be plane polarised. In the laboratory, polarisation of microwaves can be demonstrated by using a **metal grille** and polarisation of radio waves by using an **aerial**.

The microwave transmitters used in school and college laboratories transmit plane-polarised waves. When the metal rods of the grille are in the same direction as the oscillating electric field of the incident microwaves, the intensity of the transmitted microwaves is virtually zero. Rotating the metal grille through 90° allows the microwaves to pass through without much absorption. (For microwaves, the metal rods of the grille behave in the same way as the long molecular chains of a Polaroid do with light.)

The polarisation of radio waves can be demonstrated by rotating the long aerial of a radio receiver or the set-top aerial of a television.

Examples of polarised light

Examples of polarised light include:

- Light from the LCD of a laptop or a calculator is plane polarised.
- Sunlight scattered by the atmosphere is *partially* polarised.
- Reflected light from shiny surfaces, such as glass or water, is *partially* polarised.
- The light transmitted through stressed glass or plastic (e.g. a plastic ruler) is *partially* polarised.

You could test these examples for yourself using Polaroid sunglasses.

Malus's law

Plane-polarised light is incident on a Polaroid. How does the intensity of the transmitted light depend on the angle of rotation of the Polaroid?

Consider plane-polarised light of intensity I_0 incident on a Polaroid. The angle between the axis of the Polaroid and the **plane of polarisation** of the incident light is θ.

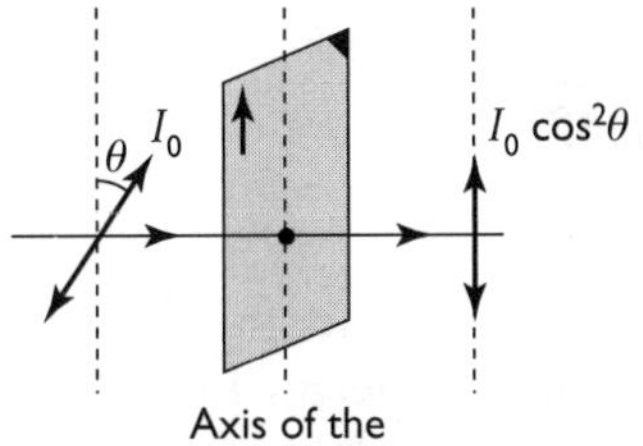

The intensity, I, of the transmitted light along the axis of the Polaroid is given by Malus's law:

$$I = I_0 \cos^2 \theta$$

The same expression can be used for plane-polarised microwaves incident at a metal grille or plane-polarised radio waves incident at an aerial.

Tip The $\cos^2\theta$ term in the expression $I = I_0 \cos^2\theta$ means $(\cos\theta)^2$.

Worked example

(a) Vertically polarised light is incident on a Polaroid, the axis of which is at 45° to the vertical. Calculate the percentage of the intensity of incident light that is transmitted through the Polaroid. Suggest what happens to light that is not transmitted.

(b) Sketch a graph of intensity of transmitted light from the Polaroid against the angle θ, which is between the vertical and the axis of the Polaroid.

Answer

(a) $I = I_0 \cos^2\theta$

The fraction of the incident intensity of light that is transmitted through the Polaroid is

$$\frac{I}{I_0} = \cos^2\theta$$

Therefore, percentage of transmitted light = $(\cos^2 45°) \times 100\% = 50\%$

The light that is not transmitted is absorbed by the long molecular chains of the Polaroid and turned into heat.

(b) The intensity of the transmitted light from the Polaroid is given by Malus's law $I = I_0 \cos^2\theta$. Hence the graph is as follows:

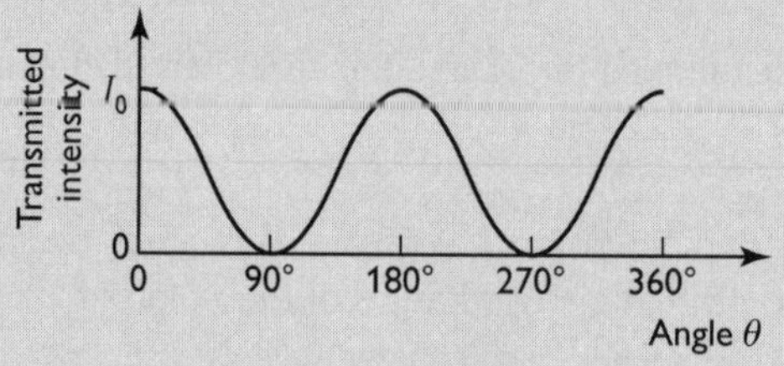

Tip The intensity is never negative because of the $\cos^2\theta$ term in Malus's law.

Interference

Superposition of waves

When two or more waves meet at a point, they pass through each other. The combining effect of the waves is called **superposition**. The principle of superposition of waves is used to determine the net displacement at a point.

The principle of superposition states that when two or more waves meet at a point, the net displacement at that point is equal to the algebraic sum of the individual displacements due to the waves.

Coherent sources

Two wave sources are **coherent** if they emit waves that have a **constant phase difference**. Hence coherent sources must emit waves of the same wavelength and frequency. Examples of coherent sources include:

- two loudspeakers connected to the same signal generator emitting sound
- two narrow slits diffracting light from a laser
- two narrow gaps diffracting microwaves from a transmitter

Two lamps connected to the same power supply do not emit light with a constant phase difference; hence they are not coherent sources.

Interference of light

Interference is term used to describe the superposition of waves from two coherent sources. The diagram below shows the diffraction of laser light from two closely spaced narrow slits.

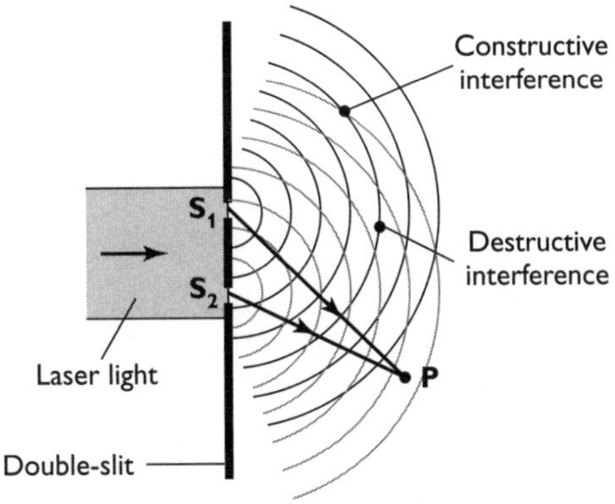

The diffracted light from the two slits combines in the space beyond the slits — they **interfere**. At some points, the waves from the slits combine **constructively** and produce a bright patch of light or fringe. At other points, the waves combine **destructively** to produce a dark fringe. Whether the waves combine constructively or destructively depends on the path difference:

> **The path difference is the extra distance travelled by one of the waves from the slits.**

In the diagram above, the path difference is $\mathbf{S_1P - S_2P}$.

If the path difference is a whole number of wavelengths, the waves arrive in phase at a point and hence produce constructive interference. If the path difference is an odd number of half wavelengths, the waves arrive with a phase difference of 180° and hence produce destructive interference. The table below is a summary of the interference effects produced by light.

Path difference	Phase difference	Type of interference	Fringe
$0, \lambda, 2\lambda \ldots n\lambda$	0°	Constructive	Bright
$\lambda/2, 1.5\lambda, 2.5\lambda \ldots (n + \frac{1}{2})\lambda$	180°	Destructive	Dark

You can observe the same effects with microwaves directed at two slits in a metal plate. The bright and dark fringes are replaced by *maximum* and *minimum* signals registered by a microwave detector moved in the region beyond the slits. The physical principles remain the same:

- The slits act as two coherent sources.
- The slits diffract the microwaves.
- The microwaves interfere constructively or destructively in the space beyond the slits.

Worked example

A microwave transmitter is directed towards two closely spaced narrow slits. The microwaves have a frequency of 12 GHz. State the type of interference (constructive or destructive) observed at a point **X** that is 75 cm from one slit and 85 cm from the other slit. Explain your answer.

Answer

The wavelength of the microwaves can be determined using the wave equation $c = f\lambda$.

$$\text{wavelength, } \lambda = \frac{c}{f} = \frac{3.0 \times 10^8}{12 \times 10^9} = 0.025\,\text{m or } 2.5\,\text{cm}$$

The path difference of the waves from the two slits = 85 − 75 = 10 cm.
The path difference is equal to four whole wavelengths, 4λ.
Hence constructive interference will be observed at point **X**.

Young's double-slit experiment

The interference of light was first demonstrated by Thomas Young in 1801.

The same effect can be shown by directing a laser beam towards two closely spaced double-slits, as shown in the diagram below.

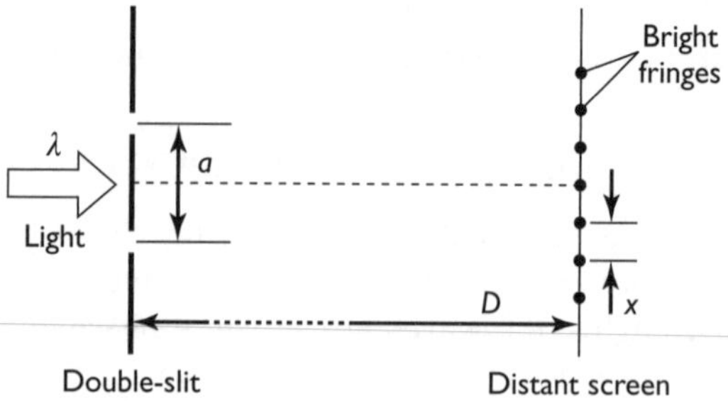

Laser light is intense, so you do not need absolute darkness to observe the interference effects. The light from a laser is also **monochromatic** (one colour) and hence the same wavelength. The incident light from the laser is diffracted at the slits. The diffracted light interferes in the space beyond the slits. A series of dark and bright interference fringes are seen on a screen placed some distance away from the double-slits. The wavelength, λ, of the monochromatic light is determined using the following equation:

$$\lambda = \frac{ax}{D}$$

where a is the separation between the slits, x is the separation between two adjacent bright (or dark) fringes and D is the distance between the slits and the screen.

The equation $\lambda = \frac{ax}{D}$ can also be used to determine the wavelength of sound waves, microwaves and radio waves. An experiment with sound would involve two loudspeakers connected to the same signal generator with the minimal and maximal signals monitored on an oscilloscope using a microphone.

The worked example below shows that the wavelength of laser light can be determined easily in the laboratory using a metre rule.

Worked example

A laser beam is directed at right angles to two narrow slits with a separation of 0.25 mm. The laser produces monochromatic red light. Dark and bright red fringes are observed on a screen placed at a distance of 3.5 m from the slits. The distance separating the extremes of 13 bright fringes is measured with a ruler and found to be 11.0 cm.

(a) Determine the wavelength of the laser light.

(b) Describe what would happen to the interference pattern if the screen were brought closer to the slits.

Answer

(a) With 13 fringes, there are 12 'separations'. Therefore:

separation between adjacent fringes $= \frac{11.0}{12} = 0.91666\ldots \text{ cm} \approx 0.92 \text{ cm}$

$$\lambda = \frac{ax}{D} = \frac{0.25 \times 10^{-3} \times 0.91666\ldots \times 10^{-2}}{3.5}$$

$$\lambda = 6.548 \times 10^{-7} \text{ m} \approx 6.5 \times 10^{-7} \text{ m}$$

The wavelength of the light from the laser is about 6.5×10^{-7} m (or 650 nm).

(b) The separation between the adjacent bright fringes is given by:

$$x = \frac{\lambda D}{a} \propto D$$

Since x is directly proportional to D, the separation between adjacent fringes decreases. Hence the interference pattern becomes more compact.

Diffraction grating

A diffraction grating consists of a large number of **lines** ruled on a glass or plastic slide. There can be as many as 600 lines every millimetre of the grating. Each line behaves as a slit, diffracting the incident light. A diffraction grating can be used to observe spectra and to determine the wavelength of visible light. The advantages of using a diffraction grating over double slits are that:

- the interference fringes are bright and sharp
- the angles between the direction of the incident light and the fringes are large, which improves the accuracy of the experiment

The diagram below shows monochromatic light of wavelength λ incident at right angles to a diffraction grating.

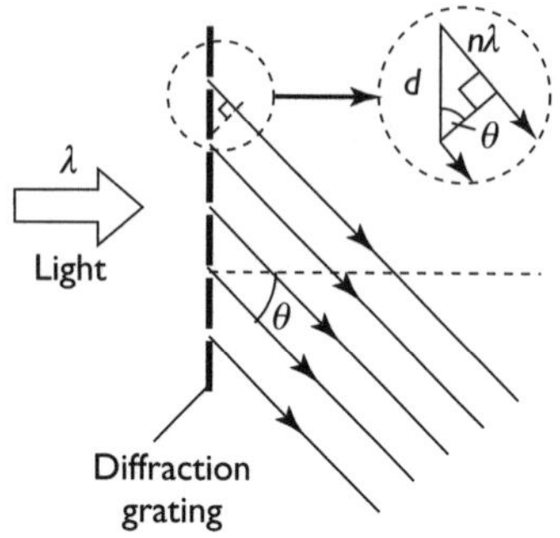

The separation between adjacent lines of the grating is d. This separation is known as the **grating spacing** or the **grating element**. In the straight forward direction, the angle θ is zero and all the waves from the slits are in phase. In this direction, constructive interference is observed. Constructive interference also occurs at other angles. For two adjacent rays, the path difference must be a whole number of wavelengths:

path difference = $d \sin\theta$ = whole number of wavelengths

Therefore:

$$\boldsymbol{d \sin\theta = n\lambda}$$

where n is known as the **order** of the maximum. (The maximum is a term used to denote the bright interference spot observed on the screen.) The value of n can only be an integer, i.e. 0, ±1, ±2, ±3 and so on.

Worked example

Intense white light from a filament lamp is incident normally at a diffraction grating with 1000 lines per millimetre. This produces several continuous spectral bands on a distant screen. For the first-order image, calculate the angle between the red and violet ends of the spectrum. Visible light has a wavelength range of 4.0×10^{-7} m to 7.0×10^{-7} m.

Answer

The separation between adjacent lines is $d = \dfrac{1\times10^{-3}}{1000} = 1.0 \times 10^{-6}$ m

For red light, we have:

$d\sin\theta = n\lambda$ with $n = 1$, $\lambda = 7.0 \times 10^{-7}$ m

$\sin\theta = \dfrac{n\lambda}{d} = \dfrac{1\times7.0\times10^{-7}}{1.0\times10^{-6}} = 0.700$ and $\theta = 44.4°$

For violet light, we have:

$d\sin\theta = n\lambda$ with $n = 1$, $\lambda = 4.0 \times 10^{-7}$ m

$\sin\theta = \dfrac{n\lambda}{d} = \dfrac{1\times4.0\times10^{-7}}{1.0\times10^{-6}} = 0.40$ and $\theta = 23.6°$

Angular separation between the red and violet ends of the first-order spectrum is:

$$\text{angle} = 44.4° - 23.6°$$
$$= 20.8° \approx 21°$$

Stationary waves

Formation of stationary waves

A **stationary wave** is also known as a **standing wave**. It is formed when two identical progressive waves of the same amplitude travelling in *opposite* directions combine. In the laboratory, this is possible by the superposition of an incident and a reflected wave. The progressive waves have the same wavelength, λ, and speed, v.

All waves can produce a standing wave pattern. A standing wave pattern is recognisable by its **nodes** and **antinodes**.

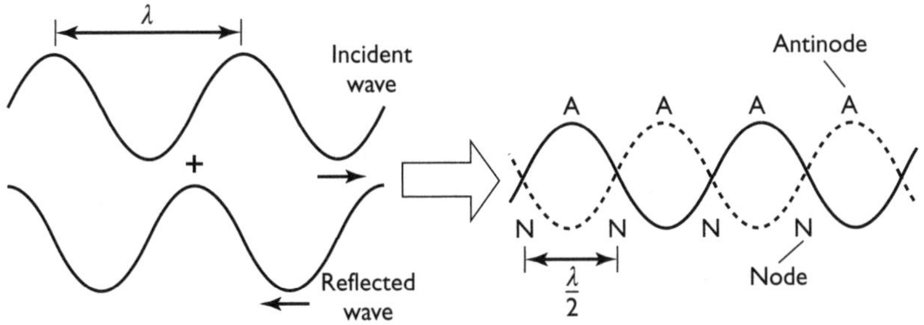

A node is a point that has zero amplitude. An antinode is a point that has maximum amplitude.

Some important points about a stationary wave include:

- The separation between neighbouring nodes (or antinodes) is equal to $\frac{\lambda}{2}$, where λ is the wavelength of the progressive wave.
- The node–antinode separation is $\frac{\lambda}{4}$.
- The stationary wave has the same frequency, f, as the progressive wave.
- We can apply the wave equation $v = f\lambda$ to determine the speed, v, of the progressive waves forming the stationary wave.
- The wave speed of the stationary wave is zero. Therefore, a stationary wave does not transfer energy.
- All points between adjacent nodes oscillate in phase — that is, the phase difference is zero.
- A point within one 'loop' and another point in the neighbouring 'loop' have a phase difference of 180°.

Stationary waves on a stretched string

Stationary waves can be produced on a stretched string by fixing one end and attaching the other end to a mechanical oscillator. Stationary wave patterns, identified by an integer number of 'loops', are formed at certain frequencies of the oscillator.

The diagram below illustrates some of the characteristics of the stationary wave patterns formed on a stretched string.

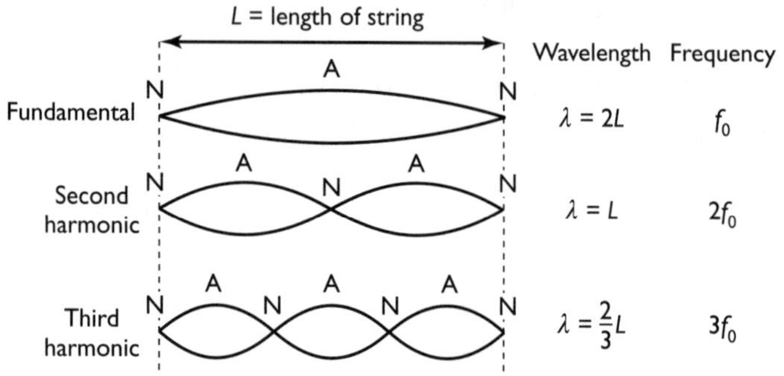

The frequency of the stationary waves is quantised — it is a multiple of the fundamental frequency, f_0.

You can also get a standing wave with one antinode by plucking the string in the middle. This vibration of the string is called its **fundamental mode of vibration**. The **fundamental frequency** f_0 is the *minimum* frequency of a stationary wave for a given arrangement.

Stationary waves in air columns

The diagram below shows how a stationary wave can be produced in an air column.

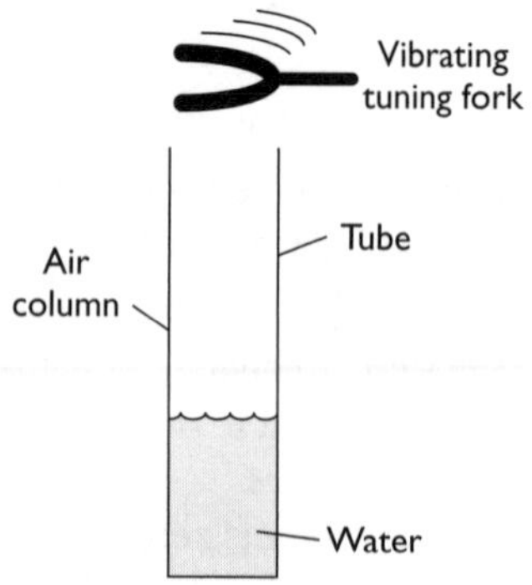

The stationary wave is the result of the superposition of the incident sound wave created by the tuning fork (or loudspeaker) and the sound reflected off the water surface. A stationary wave produces a loud 'resonating' sound from the tube.

The diagram below illustrates some of the characteristics of the stationary wave patterns formed in a pipe that is closed at one end.

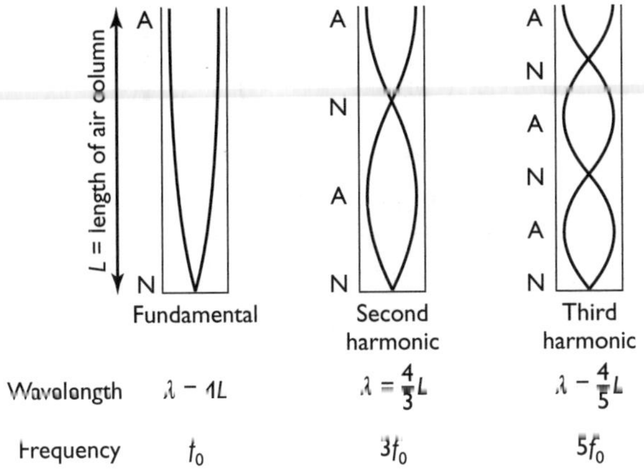

The frequency of the stationary wave is also quantised — it is an odd multiple of the fundamental frequency, f_0.

The arrangement shown in the diagram above can be used to determine the speed of sound, as is illustrated in the worked example below.

Worked example

A plastic tube is partly filled with water. A loud sound is produced from the tube when a vibrating tuning fork of frequency 440 Hz is placed at the top of the tube. The length of the air column is 19.3 cm and the frequency of the vibration is known to be fundamental. Determine the speed of sound.

Answer

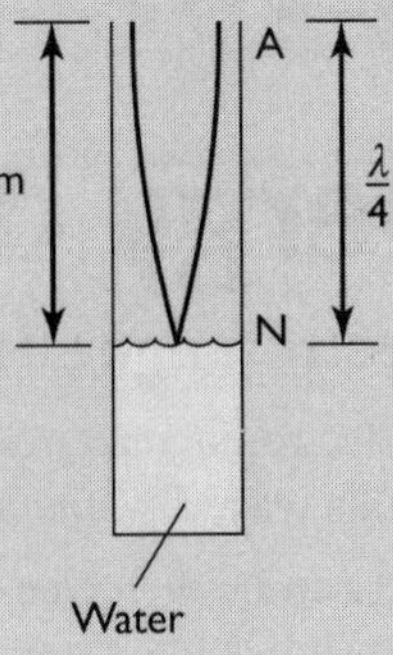

The length of the air column $= \frac{\lambda}{4}$

$$\frac{\lambda}{4} = 0.193$$

$$\lambda = 4 \times 0.193 = 0.772\ \text{m}$$

The wavelength of the sound from the tuning fork is equal to 0.772 m. The speed of sound v is given by the wave equation:

$$v = f\lambda$$

$$v = 440 \times 0.772 = 340\ \text{m s}^{-1}$$

The speed of sound is $340\ \text{m s}^{-1}$.

Stationary waves with microwaves

The diagram below shows an arrangement that can be used to demonstrate stationary waves using microwaves.

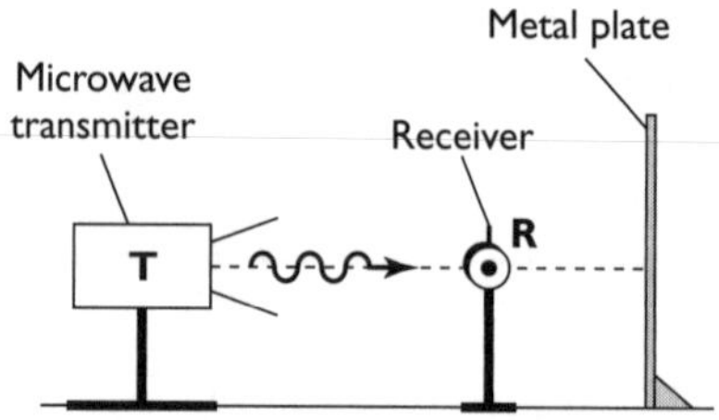

The waves from the transmitter **T** are reflected by the metal plate. The superposition of the waves from the transmitter and the reflected waves from the metal plate produces a stationary wave in the space between the transmitter and the plate. The positions of the nodes, or the antinodes, are located using the receiver **R**.

Worked example

A 12 GHz microwave transmitter is placed in front of a large metal plate. A receiver is moved towards the plate along a line at right angles to the plate. The receiver detects nine consecutive antinodes over a distance of 10.0 cm. Determine the speed of the microwaves.

Answer

The separation between adjacent antinodes = $\frac{\lambda}{2}$.

With nine antinodes, there are eight 'separations'. Therefore: $8 \times \frac{\lambda}{2} = 10.0\,\text{cm}$

$$\text{wavelength of microwaves} = \frac{0.100 \times 2}{8} = 0.025\,\text{m}$$

The speed v of the microwaves can be calculated using the wave equation. Therefore:

$$v = f\lambda = 12 \times 10^9 \times 0.025$$
$$v = 3.0 \times 10^8\,\text{m}\,\text{s}^{-1}$$

Tip It is important to convert the frequency from GHz to Hz (1 GHz = 10^9 Hz).

Energy of a photon

Electromagnetic waves: waves or particles?

All electromagnetic radiation (e.g. light) travels through space as **waves**. The evidence for this wave-like behaviour is that:

- it can be *diffracted*
- it shows *interference* effects

From p. 44 we also know that waves transmit energy at a rate that depends on their amplitude (intensity ∝ amplitude^2).

At the beginning of the twentieth century, the wave properties of electromagnetic waves did not explain how they interact on the atomic level with electrons and atoms. This interaction was explained by the invention of a radical way of interpreting electromagnetic waves: the energy of electromagnetic waves is carried through space by **photons**. The speed of the photons is $3.0 \times 10^8\,\mathrm{m\,s^{-1}}$.

Electromagnetic waves show both particle-like and wave-like behaviour. This dual nature is referred to as **wave–particle duality**. The wave–particle duality of electromagnetic waves can be summarised as follows:

- Electromagnetic waves *travel* through space as *waves*.
- Electromagnetic waves *interact* with matter as *photons*.

The particle-like behaviour of electromagnetic waves can be observed in the laboratory by placing a Geiger counter in front of a radioactive source emitting gamma rays. The counter registers individual 'clicks'. Each click is the physical detection of a single gamma-ray photon.

Photons

Two eminent physicists, Max Planck and Albert Einstein, introduced the concept of photons.

> **A photon is a packet of electromagnetic energy or a quantum of electromagnetic energy.**

The energy E of a photon is related to the frequency f of the electromagnetic wave by the following equation:

$$E = hf$$

where h is the Planck constant — an important constant in quantum physics. It has an experimental value of $6.63 \times 10^{-34}\,\mathrm{J\,s}$.

The speed of an electromagnetic wave is given by the wave equation $c = f\lambda$, where λ is the wavelength of the wave and c is the speed of light in a vacuum. Hence, the energy of a photon can be calculated using the following relation:

$$E = \frac{hc}{\lambda}$$

The following points are worth noting:

- The energy of a photon is *directly* proportional to the frequency and *inversely* proportional to the wavelength of the electromagnetic wave. Higher frequency or shorter wavelength implies more energetic photons. For example, a photon of gamma rays is much more energetic than a photon of radio waves.
- The particle-like property of the photon (its energy) is intertwined with its wave-like property (its wavelength or frequency).

Worked example

A red LED emits monochromatic light of wavelength 670 nm. The radiant power of the LED is 20 mW. Calculate:

(a) the energy of a single photon

(b) the rate at which the photons are emitted from the LED

Answer

(a) The energy E of a single photon is:

$$E = \frac{hc}{\lambda} = \frac{6.63\times10^{-34}\times3.0\times10^{8}}{6.7\times10^{-7}}$$

$E = 2.97 \times 10^{-19}\,J \approx 3.0 \times 10^{-19}\,J$

(b) The LED emits $20 \times 10^{-3}\,J$ every second.

The rate of photons emitted is the same as the number of photons emitted every second. Hence:

$$\text{number of photons emitted per second} = \frac{\text{energy emitted every second}}{\text{energy of a single photon}}$$

$$= \frac{20\times10^{-3}}{2.97\times10^{-19}} \approx 6.7 \times 10^{16}\,s^{-1}$$

There are about 67 thousand million million photons emitted every second from the LED.

Electronvolt (eV)

The electronvolt (eV) is a convenient unit of energy for dealing with particles and photons. It is an ideal unit when the energy in joules is minuscule. It is defined as follows:

The electronvolt is the energy gained by an electron travelling through a potential difference of 1 volt.

What is 1 electronvolt (eV) in joules (J)?

work done on the electron = charge on the electron × potential difference.

$1\,eV = 1.6 \times 10^{-19}\,C \times 1\,V$ or $\mathbf{1\,eV = 1.6 \times 10^{-19}\,J}$

The electronvolt is an easy energy unit to use. An electron travelling through a potential difference of 1000 V gains 1000 eV of energy. Similarly, an electron or a proton (same *numerical* charge as the electron) accelerated in a particle accelerator through a potential difference of 1 million volts gains an energy of 1 MeV.

Here are a couple of helpful rules:

- To convert from electronvolts into joules, multiply by 1.6×10^{-19}.
- To convert from joules into electronvolts, divide by 1.6×10^{-19}.

Worked example

(a) Estimate the energy of a photon of gamma rays in joules and in electronvolts.

(b) Use your answer to **(a)** to estimate the potential difference through which an electron has to travel to acquire the same energy as the gamma-ray photon.

Answer

(a) The wavelength of gamma rays lies in the range 10^{-13} m to 10^{-16} m. The estimation is based on a wavelength of 10^{-15} m.

Energy of photon, $E = \frac{hc}{\lambda}$

In joules: $E = \frac{6.63 \times 10^{-34} \times 3.0 \times 10^{8}}{10^{-15}}$

$= 1.99 \times 10^{-10}\,\text{J} \approx 2.0 \times 10^{-10}\,\text{J}$

In electronvolts: $E = \frac{1.99 \times 10^{-10}}{1.6 \times 10^{-19}} \approx 1.2 \times 10^{6}\,\text{eV}$

(b) An electron has to be accelerated through a potential difference of 1.2 million volts in order to have the same energy as a 1.2 MeV photon.

Photons and light-emitting diodes

Like all diodes, an LED conducts in only one direction. An LED is designed to emit light when it conducts and has a minimum forward voltage (known as the **threshold voltage**) before it starts to conduct. Different coloured LEDs have different threshold voltages.

What happens inside the LED when it starts to emit light? The theory outlined below is simple and crude, but it can be used to make an estimate of the value of the Planck constant.

At the threshold voltage, V, the electrical energy lost by a single electron reappears as the energy of a single photon of light:

energy lost by electron = energy of a photon

$$QV = \frac{hc}{\lambda}$$

For an electron, the charge $Q = e = 1.6 \times 10^{-19}$ C.

Therefore:

$$\boldsymbol{eV = \frac{hc}{\lambda}}$$

Rearranging gives:

$$V = \left(\frac{hc}{e}\right) \times \frac{1}{\lambda}$$

Compare this with the equation for a straight line through the origin, $y = mx$.

By plotting V on the y-axis and $\frac{1}{\lambda}$ on the x-axis, the gradient of the straight line through the origin is $\frac{hc}{e}$. Therefore, a value for Planck constant can be estimated by plotting a graph as shown on p. 60.

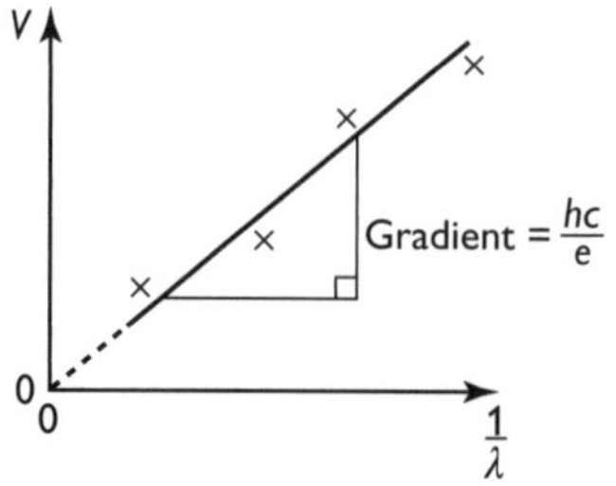

Worked example

Use the data in the table below to estimate a value for the Planck constant, h.

Colour of LED	Threshold voltage, V/V	Wavelength, $\lambda/10^{-7}$ m
Red	1.4	6.7
Amber	2.0	6.1
Green	2.3	5.6

Answer

energy lost by electron = energy of a photon

$$eV = \frac{hc}{\lambda}$$

$$h = \frac{eV\lambda}{c}$$

For the red LED, we have:

$$h = \frac{1.6\times10^{-19}\times1.4\times6.7\times10^{-7}}{3.0\times10^{8}} = 5.0 \times 10^{-34}\,\text{J s}$$

The table shows the other values calculated for h.

Colour of LED	Threshold voltage, V/V	Wavelength, $\lambda/10^{-7}$ m	$h = \frac{eV\lambda}{c}/10^{-34}$ J s
Red	1.4	6.7	5.0
Amber	2.0	6.1	6.5
Green	2.3	5.6	6.9

The average value for the Planck constant is about 6.1×10^{-34} J s.

The photoelectric effect

Understanding the photoelectric effect

The phenomenon of the removal of electrons from the surface of a metal using electromagnetic radiation is the **photoelectric effect**. The electrons removed in this way are often referred to as **photoelectrons**.

Some of the negatively charged electrons in a metal are loosely attached to the positive ions. External energy can be supplied to remove these electrons. The work function ϕ of the metal is defined as follows:

The work function of a metal is the minimum energy required to free an electron from the surface of the metal.

The diagram below shows a simple experiment that may be used to demonstrate the photoelectric effect.

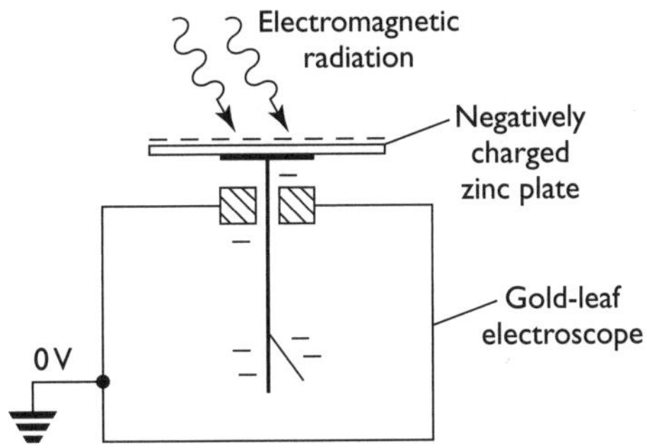

The zinc plate is attached to a gold-leaf electroscope and is given an excess of electrons by momentarily connecting it to the negative terminal of a 5.0 kV supply. The gold leaf of the electroscope diverges because of the repulsion between the electrons.

What happens when the plate is exposed to electromagnetic radiation? Shining a filament lamp on the plate does not make any difference to the divergence of the gold leaf. The visible light from the lamp does not remove a single electron from the metal surface. They remain embedded within the metal even when the intensity of the incident light is greatly increased. Therefore, the conclusion must be that light does not interact with the electrons as waves.

Using ultraviolet radiation from a mercury discharge lamp, electrons are removed *instantaneously*. The gold leaf falls quickly as the zinc plate loses electrons. Increasing the intensity of the ultraviolet radiation increases the rate at which the electrons are removed from the metal surface. Electrons are removed even when the intensity of the incident ultraviolet radiation is feeble.

Ultraviolet radiation has a higher frequency than visible light. Why should this make a difference? The removal of electrons from the metal surface can be explained only in terms of photons interacting with the electrons. An electron can only be removed from the metal surface when the energy of a single photon is equal to or greater than the work function ϕ of the metal. The photon energy of visible light is less than ϕ; the photon energy of ultraviolet radiation is greater than ϕ.

Here are some helpful rules for the photoelectric effect:

- A single photon interacts with a single electron — it is a one-to-one interaction.
- Photons interact with the surface electrons of the metal.

- Energy must be conserved in the interaction between a photon and an electron.
- Electrons are only removed when the energy of a photon is greater than or equal to the work function of the metal:

  $hf \geq \phi$
- The intensity of the incident radiation is related to the rate at which the photons are incident on the metal surface. The incident rate of the photons is doubled when the intensity of the wave is doubled.

Worked example

The work function of gold is 4.90 eV.

(a) Calculate the work function of gold in joules.

(b) A gold plate is exposed to ultraviolet radiation of wavelength 2.80×10^{-7} m. Explain why there is no photoelectric effect, even when the intensity of the radiation is doubled.

(c) Calculate the minimum frequency of the incident electromagnetic waves that will produce photoelectrons.

Answer

(a) $1\,\text{eV} = 1.6 \times 10^{-19}\,\text{J}$

Therefore, $\phi = 4.90 \times (1.6 \times 10^{-19}) = 7.84 \times 10^{-19}\,\text{J}$

(b) The energy of a single photon is:

$$E = \frac{hc}{\lambda} = \frac{6.63\times10^{-34}\times3.0\times10^{8}}{2.8\times10^{-7}} = 7.10 \times 10^{-19}\,\text{J}$$

The energy of a single photon is less than the work function ϕ of the metal. Hence the electrons cannot be removed from the gold plate by the incident waves.

Increasing the intensity increases only the rate of the incident photons. The energy of the each photon is unchanged and still less than ϕ. Hence the electrons remain embedded within the plate.

(c) When the photoelectron is just removed from the metal surface, we have:

energy of photon = work function of metal

$hf_0 = \phi$

$$f_0 = \frac{7.84\times10^{-19}}{6.63\times10^{-34}} = 1.18 \times 10^{15}\,\text{Hz}$$

The minimum frequency, f_0, is 1.18×10^{15} Hz. Photoelectrons are not ejected when the frequency of the incident waves is less than f_0.

Einstein's photoelectric equation

Imagine a single photon interacting with a single surface electron. In 1905, Einstein applied the principle of conservation of energy to this interaction and came up with this equation for the photoelectric effect:

energy of photon = work function + maximum kinetic energy of photoelectron

$$hf = \phi + \mathrm{KE}_{max}$$

This equation is called Einstein's photoelectric equation.

Energy of the photon is less than ϕ

Not a single electron can be removed from the metal surface when $hf < \phi$. This remains the case even when the intensity of the incident waves is increased. An electron interacting with a photon remains within the metal. The electron acquires some kinetic energy that is dispersed quickly as heat to the rest of the metal.

Energy of the photon is equal to ϕ

The photoelectron is *just* removed from the metal surface. The kinetic energy of the photoelectron is zero. The frequency of the incident waves is the minimum that will eject photoelectrons from the metal surface. This minimum frequency is known as the threshold frequency f_0. It is defined as follows:

The threshold frequency is the minimum frequency of the incident electromagnetic waves that will produce photoelectrons.

Therefore:

$$hf_0 = \phi$$

or

$$\text{threshold frequency } f_0 = \frac{\phi}{h}$$

Note: No photoelectrons can be emitted if the frequency of the incident waves is less than f_0.

Energy of the photon is greater than ϕ

Photoelectrons are *instantaneously* emitted from the metal surface when the incident electromagnetic waves have a frequency greater than the threshold frequency. The electrons emerge with a range of kinetic energies. Those with the least kinetic energy are those that are firmly attached to the metal ions. The electrons with the greatest kinetic energy are the electrons in the metal that are loosely attached. The maximum kinetic energy KE_{max} of the electrons is given by:

$$\mathrm{KE}_{max} = hf - \phi$$

A graph of KE_{max} against frequency, f, of the incident electromagnetic waves is a straight line.

The equation $\mathrm{KE}_{max} = hf - \phi$ can be compared with the equation for a straight line, $y = mx + c$. The following conclusions can be drawn from this graph:

- The gradient of the line is the Planck constant, h. The slope of the line is independent of the type of metal used.

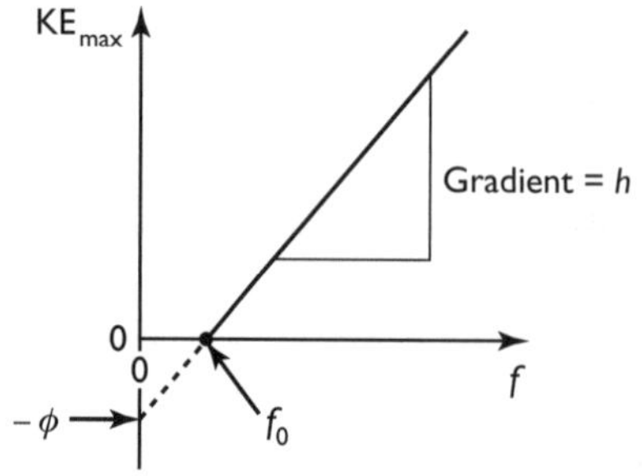

- The intercept with the KE_{max} -axis is equal to $-\phi$.
- The intercept with the f-axis is equal to the threshold frequency, f_0.

Worked example

A negatively charged zinc plate is exposed to electromagnetic radiation. The work function of zinc is 6.9×10^{-19} J.

(a) Calculate the maximum kinetic energy in electronvolts of the electrons when the incident waves have a frequency of 2.5×10^{15} Hz.

(b) State and explain how your answer to **(a)** would change, if at all, when the intensity of the incident radiation is doubled.

Answer

(a) Using Einstein's photoelectric equation, we have:

$hf = \phi + KE_{max}$

$KE_{max} = hf - \phi = (6.63 \times 10^{-34} \times 2.5 \times 10^{15}) - 6.9 \times 10^{-19}$

$KE_{max} = 9.675 \times 10^{-19}\,J \approx 9.68 \times 10^{-19}\,J$

$$KE_{max} = \frac{9.675 \times 10^{-19}}{1.6 \times 10^{-19}} = 6.047\,eV \approx 6.0\,eV$$

(b) Doubling the intensity of the incident radiation doubles the rate of photons incident on the metal plate. The energy of each photon is not affected by intensity. Hence, the maximum kinetic energy of the electrons remains the same.

Wave–particle duality

Particles: particles or waves?

All particles have mass. Some — for example the proton and the electron — have a charge. All particles are accelerated by forces. We can use Newtonian mechanics (e.g. $F = ma$) to describe the motion of particles.

On p. 57 we saw that electromagnetic waves have a dual nature. They travel as waves but interact as 'particles' (photons). In 1923, Prince Louis de Broglie wondered if particles such as electrons also have a dual nature. Electrons interact with matter as particles, but do they travel through space as a wave?

De Broglie proposed that a particle of mass m travelling at a speed v can be represented by a wave of wavelength λ, as shown by the following equation (known as the de Broglie equation):

$$\lambda = \frac{h}{mv}$$

where h is the Planck constant. The product mv is the momentum p of the particle. Hence the de Broglie equation can also be written as:

$$\lambda = \frac{h}{p}$$

The waves associated with moving particles are sometimes referred to as **matter waves** or simply **de Broglie waves**. These waves are *not* electromagnetic waves.

Tip Referring to de Broglie waves as electromagnetic is a common mistake made by candidates in exams. These waves have peculiar properties and detailed knowledge is not required for this specification.

If electrons travel through space as waves, then they should show diffraction effects. The wave-like behaviour of electrons can be demonstrated using a diffraction tube, as shown in the diagram below.

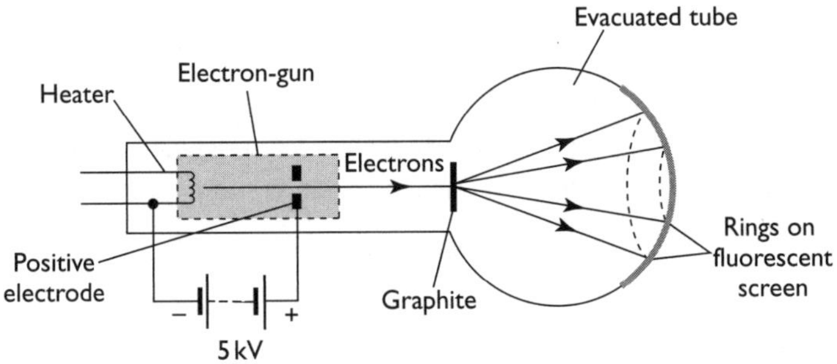

- The electrons are 'boiled' off the heater, which is a small metal filament.
- These electrons are accelerated by the large voltage at the anode (typically 5 kV).
- The electrons travel through a thin layer of graphite.
- The electrons are diffracted by the graphite and show 'diffraction rings' on the fluorescent screen.

Diffraction effects can be seen because the electrons have a de Broglie wavelength that is similar to the diameter of the graphite atoms and to the spacing between the atomic layers, which is of the order of 10^{-10} m. The spaces between the layers of atoms behave like a diffraction grating to the electron waves.

Worked example

Electrons in a diffraction tube are accelerated through a voltage of 5.0 kV. The electrons pass through a thin layer of graphite. The spacing between atomic layers in graphite is about 3×10^{-10} m.

(a) Show that the speed v of the electrons is about $10^7\,m\,s^{-1}$.
(b) Calculate the de Broglie wavelength of the electrons.
(c) Explain whether or not the electrons travelling through graphite will be diffracted.

Answer

(a) The kinetic energy of an electron is 5.0 keV.
The energy has to be converted into joules.

$KE_{max} = 5.0\,keV = 5.0 \times 10^3 \times 1.6 \times 10^{-19} = 8.0 \times 10^{-16}\,J$ $(1\,eV = 1.6 \times 10^{-19}\,J)$

$KE_{max} = \frac{1}{2} mv^2$

The mass of electron is 9.11×10^{-31} kg (given on exam *Data Sheet*)
Therefore:

$$8.0 \times 10^{-16} = \frac{1}{2} \times 9.11 \times 10^{-31} \times v^2$$

$$v = \sqrt{\frac{2 \times 8.0 \times 10^{-16}}{9.11 \times 10^{-31}}} = 4.19 \times 10^7\,m\,s^{-1} \approx 4.2 \times 10^7\,m\,s^{-1}$$

(b) The wavelength of the electrons can be calculated from the de Broglie equation:

$$\lambda = \frac{h}{mv}$$

$$\lambda = \frac{6.63 \times 10^{-34}}{9.11 \times 10^{-31} \times 4.19 \times 10^7} = 1.74 \times 10^{-11}\,m \approx 1.7 \times 10^{-11}\,m$$

(c) The wavelength of the electrons is similar to the spacing between the graphite atoms. The electrons will therefore be diffracted by the graphite atoms.

People waves

The de Broglie equation can be applied to all matter — to anything that has mass and which moves. This includes people. However, people do not get diffracted when moving around. Why this is so is illustrated below.

The mass of a typical person is 70 kg. The de Broglie wavelength, λ, of a person running at a speed of $5.0\,m\,s^{-1}$ is:

$$\lambda = \frac{h}{mv}$$

$$= \frac{6.63 \times 10^{-34}}{70 \times 5.0} \approx 2 \times 10^{-36}\,m$$

This wavelength is extremely small compared with the 'gaps' we encounter. This is why we are not diffracted when running down corridors or through open doors.

Probing matter

X-rays can be diffracted by matter. This is because X-rays have wavelengths ($\sim 10^{-10}$ m) similar to the diameter of the atoms and the spacing between the atoms. The diffraction pattern can be used to deduce the arrangement of the atoms within a substance.

The wave properties of particles such as electrons and neutrons can also be used to investigate matter. Moving particles can be diffracted to produce diffraction patterns, just as X-rays can.

- **Slow-moving electrons** (10^7 m s^{-1}) have a de Broglie wavelength of about 10^{-10} m. They can be used to determine the diameter of atoms and the structure of complex molecules such as DNA. The same determination can be carried out with thermal neutrons (10^3 m s^{-1}) from a nuclear reactor
- **Fast-moving electrons** (10^8 m s^{-1}) from particle accelerators have a de Broglie wavelength of about 10^{-15} m. They can be used to determine the diameter of atomic nuclei. These electrons can also be used to deduce the internal structure of the nuclei — for example, the arrangement of protons and neutrons.

Energy levels in atoms

Emission spectrum

The atoms of a hot gas are isolated from each other. They are far apart and hence exert negligible electrical forces on each other (except when they collide). The spectrum of light emitted from hot gas atoms consists of well-defined bright lines called **emission spectral lines**. Each bright, coloured line is associated with a specific wavelength. Atoms of different elements emit their own unique pattern of spectral lines.

The emission line spectrum produced by hot gases provides direct evidence that the energy of the electrons in the atoms is **quantised**. Each electron in an atom can only exist in certain fixed and discrete energy states. These energy states are also known as **energy levels**. The diagram below shows two energy levels E_1 and E_2:

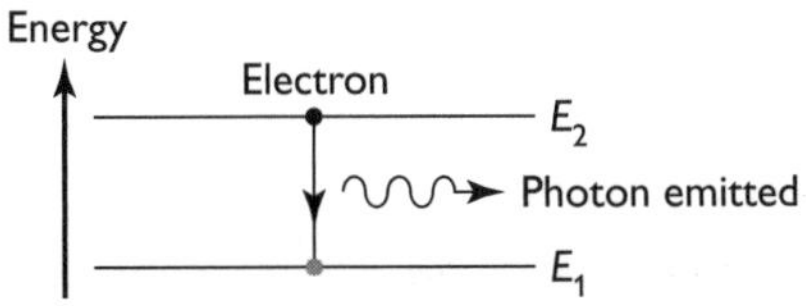

An excited electron is temporarily in the higher energy level E_2. It loses energy when it makes a transition (jump) to the lower energy level E_1. Since energy is conserved, the energy lost by the electron appears as a quantum of energy in the form of a photon. The emitted photon has a specific wavelength or frequency. An emission line spectrum is produced by large numbers of electrons making similar transitions between other pairs of energy levels.

The frequency f and the wavelength λ of the electromagnetic waves emitted can be calculated using the following equations:

$$hf = E_2 - E_1 \text{ and } \frac{hc}{\lambda} = E_2 - E_1$$

The energy levels have *negative* values. This means that external energy is required to remove electrons from the attractive forces of the positive atomic nuclei. The

ground state of the atom is the lowest energy level of the atom. The minimum energy needed to remove an electron from its lowest energy level so that it escapes from the atom is called the **ionisation energy** of the atom.

Worked example

The diagram below shows some of the energy levels of the electron in a hydrogen atom.

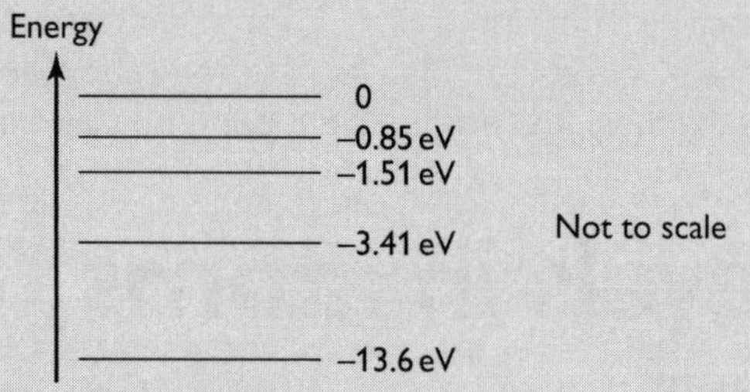

(a) Calculate the wavelength of the electromagnetic waves emitted when an electron makes a transition from an energy level of −1.51 eV to an energy level of −3.41 eV. In which region of the electromagnetic spectrum can the associated spectral line be detected?

(b) A electron makes a jump from energy level −1.51 eV to energy level −13.60 eV. State and explain whether the emitted wavelength will be smaller than, equal to, or greater than your answer to **(a)**.

Answer

(a) energy of photon = energy lost by electron

Therefore:

$$\frac{hc}{\lambda} = E_2 - E_1$$

$$\frac{hc}{\lambda} = -1.51 - (-3.41) = 1.90\,\text{eV}$$

The energy of the photon has to be converted into joules.

$$1.90\,\text{eV} = 1.90 \times 1.6 \times 10^{-19} = 3.04 \times 10^{-19}\,\text{J}$$

Hence:

$$\frac{6.63 \times 10^{-34} \times 3.0 \times 10^{8}}{\lambda} = 3.04 \times 10^{-19}$$

$$\lambda = \frac{6.63 \times 10^{-34} \times 3.0 \times 10^{8}}{3.04 \times 10^{-19}} = 6.54 \times 10^{-7}\,\text{m}$$

This wavelength is in the visible region of the electromagnetic spectrum.

(b) The energy of the photon is the difference between the two energy levels. Therefore:

$$\frac{hc}{\lambda} = -1.51 - (-13.60) = 12.09\,\text{eV}$$

The energy of the photon is greater than that of the photon in **(a)**. Hence the wavelength must be shorter than that in the answer to **(a)**.

Absorption spectrum

An emission spectrum is produced by hot gases; an absorption spectrum requires cooler gases, so that energy can be absorbed.

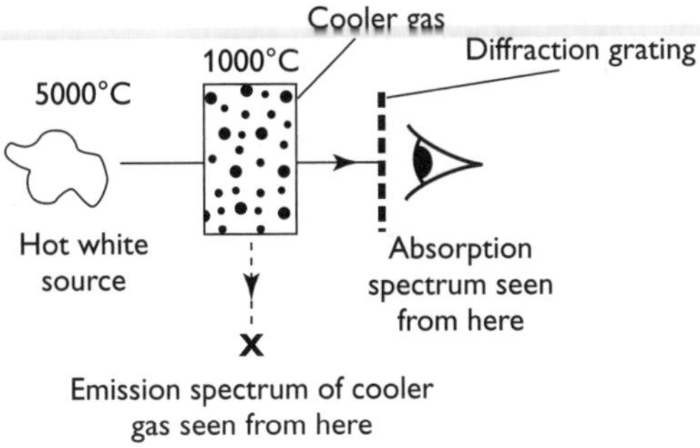

The light from the hot white source passes through a 'cooler' gas and is observed using a diffraction grating. The spectrum is continuous with sharp dark lines, called **absorption spectral lines.** These spectral lines have the same wavelengths as the emission spectral lines when observed from **X** (see the diagram above). White light consists of photons of many different energies. A photon is absorbed by an atom of the cool gas when its energy is the same as the difference between any two energy levels. If the energy is too low or too high, the photon is not absorbed. A photon absorbed by an atom can be re-emitted in a multitude of directions. This is why an absorption spectral line appears dark. The diagram below shows what happens when a photon interacts with an electron.

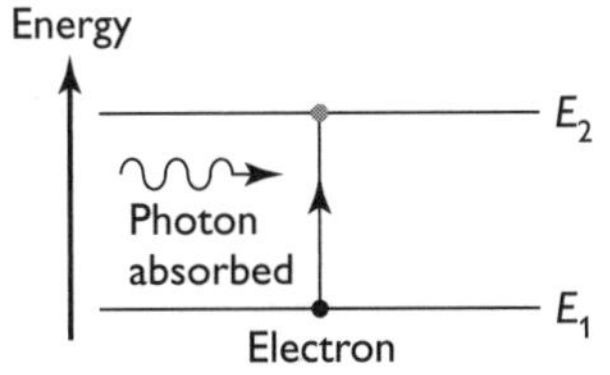

Note: If gas atoms emit monochromatic light of wavelength 450 nm, then the same atoms can also absorb light of wavelength 450 nm.

Questions & Answers

This section contains questions similar in style to those you can expect in the exam paper for Unit G482. The responses of two candidates are given. The answers from Candidate A are similar to those expected from a student who has reached grade-A standard in this unit. The answers given by Candidate B are typical of those of a grade-C candidate and may be incomplete or inappropriate. Answers from weaker candidates than this are not instructive at this stage, because there are either too many questions unanswered or the answers show a lack of knowledge of basic physics.

At least one question is given for each section of this unit.

The Question and Answer section can be used in several ways. The best way to engage your brain is to write something on a piece of paper. There is no value at all in just reading through the questions and answers given by the two candidates. Here are some possible strategies. You could:

- Attempt a question yourself and then mark it using the comments from the examiner. This will enable you to compare the responses of the two candidates and learn from your mistakes.
- Mark the answers given by the two candidates and then see whether you agree with the mark awarded by the examiner.

Examiner comments

Every mark scored by a candidate is shown by a tick at the relevant place.

For most parts, the total number of ticks (✓) and crosses (✗) should add up to the total mark for the question. All candidates' responses are followed by examiner comments, denoted by the icon *e*. These comments focus on the errors made by the candidates and sometimes offer alternative ways of securing the marks. The comments at the end of each question provide valuable tips for answering examination questions.

Question 1

Electric current

The circuit diagram below shows two resistors connected to a 12 Ω supply of negligible internal resistance.

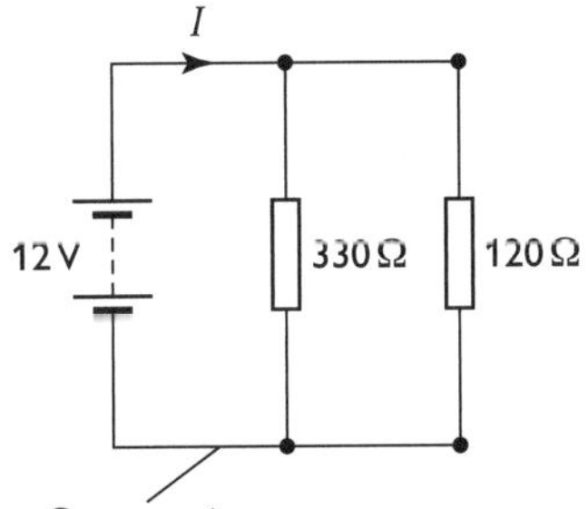

(a) Calculate the current in the resistor of resistance 120 Ω. (2 marks)

(b) State Kirchhoff's first law and use it to determine the current drawn from the supply. (3 marks)

(c) The connecting wires in the circuit are made from copper. They have negligible resistance and have a cross-sectional area of $2.5 \times 10^{-7}\,m^2$. The electron number density of copper is $8.0 \times 10^{28}\,m^{-3}$.

(i) Explain the term 'electron number density'. (1 mark)

(ii) Determine the mean drift velocity of the electrons within the copper cable. (3 marks)

Total: 9 marks

Candidates' answers to Question 1

Candidate A

(a) The current in the resistor can be found using $V = IR$:

$$I = \frac{V}{R} = \frac{12}{120} = 0.10\,\text{A} \checkmark\checkmark$$

Candidate B

(a) current = p.d./resistance ✓

current = 0.10 A = 100 mA ✓

Both candidates have made a good start. The layout of Candidate A's response is flawless. Candidate B has not shown all the working but the answer 0.10 A is correct. The candidate has also given the answer in mA when there is no need to do so.

Candidate A

(b) Kirchhoff's first law states that the total of the currents entering a point in a circuit is equal to the sum of the currents leaving that same point ✓.

The current in the 330 Ω resistor = $\frac{12}{330}$ = 0.0364 ≈ 0.036 A ✓

The total current from the supply = $I_1 + I_2$ = 0.10 + 0.036 = 0.136 A ✓

Candidate B

(b) The law is:

sum of current into a junction = sum of current exiting the same junction ✓

The current $I = \frac{12}{330} + 0.10 = 0.13636$ A ✓ ✓

The answers from the two candidates look different but the end result is the same. There was no need for Candidate B to quote the final current to so many significant figures. Since the data are given to two significant figures, the answers from both candidates should be written as 0.14 A.

Candidate A

(c) (i) This is written as *n* and is equal to the number of free electrons per cubic metre of the substance ✓. Metals have larger values for *n* than insulators.

Candidate B

(c) (I) The number density is equal to the number of conduction electrons per cubic metre of the material ✓.

Both candidates gain a mark. Candidate A correctly defines the term but also provides an unnecessary statement. This wastes time.

Candidate A

(c) (ii) $I = Anev$

$$v = \frac{I}{Ane} = \frac{0.136}{2.5\times10^{-7}\times8.0\times10^{28}\times1.6\times10^{-19}}$$ ✓ ✓

$v = 4.25 \times 10^{-5}\,\mathrm{m\,s^{-1}} \approx 4.3 \times 10^{-5}\,\mathrm{m\,s^{-1}}$ ✓

Candidate B

(c) (ii) $v = \frac{Ane}{I}$ = 23 500 metres per second ✗ ✗ ✗

Candidate A has provided yet another perfect answer. Candidate B has used the equation $I = Anev$ from the physics *Data Sheet*, but has rearranged it incorrectly. Examiners do not award marks for substitution into an incorrect equation.

Overall, Candidate A scores the full 9 marks; the layout of the answers is commendable. Candidate B scores 6 marks. Not being able to rearrange the equation in (c)(ii) has proved costly. This candidate would have gained by showing all the working; securing 1 more mark would have ensured a grade B.

Question 2

Electromotive force (e.m.f.) and potential difference (p.d.)

The circuit diagram below shows a filament lamp connected across a cell.

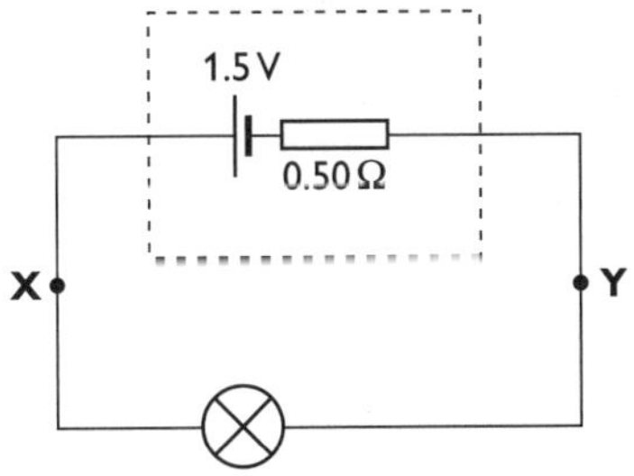

The cell has electromotive force (e.m.f.) of 1.5 V and an internal resistance of 0.50 Ω. The current drawn from the cell is 0.80 A. The potential difference (p.d.) across the lamp is 1.1 V. The switch S is closed for a duration of exactly 1 minute.

(a) State one similarity between p.d. and e.m.f. (1 mark)

(b) Calculate:

(i) the flow of charge through the lamp (2 marks)

(ii) the number of electrons passing through the lamp (2 marks)

(iii) the electrical energy transformed by the lamp (2 marks)

(iv) the power dissipated by the lamp (2 marks)

(c) A high-resistance voltmeter is placed between the points X and Y. Explain why this reading is different from the e.m.f. of the cell. (1 mark)

Total: 10 marks

Candidates' answers to Question 2

Candidate A

(a) Both are defined in terms of energy divided by charge ✓. Both have the same unit — the volt.

Candidate B

(a) Voltage and e.m.f. are measured in volts ✓.

e Both candidates have recalled the physics correctly. Candidate A has given two correct similarities, either of which would earn the mark. Candidate B has used the term voltage, which is the same as potential difference, so the answer gains the mark.

Candidate A

(b) (i) $I = 0.80\,\text{A}$; $\Delta t = 60\,\text{s}$; $\Delta Q = ?$

$\Delta Q = I\Delta t$ ✓

charge $= 0.80 \times 60 = 48\,\text{C}$ ✓

Candidate B

(b) (i) charge = current × time ✓

charge $= 0.80\,\text{A} \times 60\,\text{s} = 48\,\text{A s}$ ✓

e Both candidates gain 2 marks. Candidate B has quoted the unit for charge as 'ampere second'. Since 1 A s = 1 C, the physics is correct and the mark is awarded.

Candidate A

(b) (ii) number of electrons $= \dfrac{48}{1.6 \times 10^{-19}} = 3.0 \times 10^{20}$ ✓ ✓

Candidate B

(b) (ii) I don't know how to find the number of electrons ✗ ✗.

e Candidate B has offered no answer to this tough question. The flow of charge in the circuit is due to the flow of electrons. This important connection was not made by Candidate B.

Candidate A

(b) (iii) energy = voltage × charge ✓

energy $= 1.1 \times 48 \approx 53\,\text{J}$ ✓

The electrical energy is transformed into heat and light in the lamp.

Candidate B

(b) (iii) $V = \dfrac{W}{Q}$

energy $= W = VQ = 1.5 \times 48 = 72\,\text{J}$ ✓ ✗

e Candidates often find this type of question difficult. Candidate A's answer is good. Candidate B makes a serious error by substituting the value of the e.m.f., rather than the p.d., across the lamp. Fortunately, the candidate shows good physics and therefore gains 1 mark.

Candidate A

(b) (iv) power = energy/time

power $= \dfrac{53}{60} = 0.88\,\text{W}$ ✓ ✓

Candidate B

(b) (iv) $P = VI$

power $= 1.1 \times 0.80 = 0.88\,\text{W}$ ✓ ✓

e The candidates have approached this question differently. Candidate A has used the previous answers as a guide, whereas Candidate B opts for the equation $P = VI$.

Candidate A

(c) The voltmeter reading will be 1.1 V. This is less than the e.m.f. of 1.5 V because there is a voltage of 0.40 V across the internal resistance of the cell ✓.

Candidate B

(c) voltage across the internal resistor = $0.80 \times 0.50 = 0.40$ V ✓
The voltmeter measures the voltage across the terminals of the cell.

e Both candidates realise that the internal resistance of the cell has a part to play. Both candidates score the mark.

e **Overall, Candidate A scores the full 10 marks and typifies what an A-grade student can produce in an examination. The answers are complete and contain just the right amount of physics. Candidate B has managed to gain 7 marks. Some attempt should have been made to answer (b)(ii). Using the wrong value for the voltage in (b)(iii) cost this candidate a grade.**

Question 3

Resistance and resistivity

(a) Write a word equation for the resistivity of a material. (2 marks)

(b) Explain how the resistivity of a semiconductor changes with temperature. (2 marks)

(c) The diagram below shows a pencil lead and its dimensions.

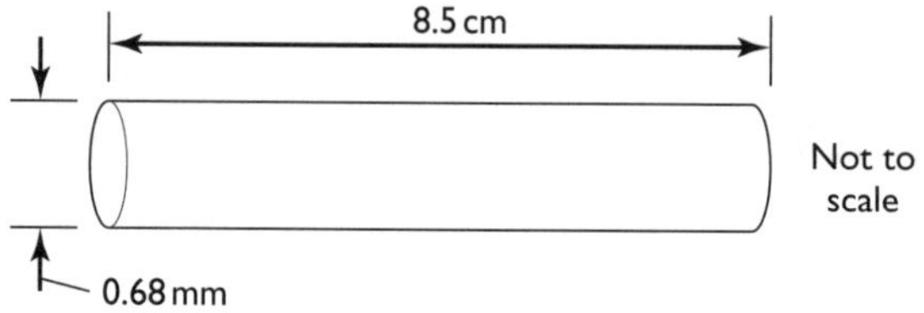

The pencil lead is 8.5 cm long and has diameter 0.68 mm. The lead is made of a material of resistivity $7.8 \times 10^{-6}\,\Omega$ m.

(i) Calculate the resistance of the pencil lead. (3 marks)

(ii) Calculate the current in the pencil lead when connected to a DC supply of e.m.f. 12 V and internal resistance 0.50 W. (3 marks)

Total: 10 marks

Candidates' answer to Question 3

Candidate A

(a) Resistivity is given by the following equation:

$$\text{resistivity} = \frac{\text{resistance} \times \text{cross-sectional area}}{\text{length}}$$ ✓✓

Candidate B

(a) The resistance equation is $R = \frac{\rho L}{A}$. Therefore:

$$\rho = \frac{RA}{L}$$ ✓

e Candidate A has written a precise word equation and taken care to make sure to include 'cross-sectional' with the 'area'. Candidate B has given the correct equation but has not defined the terms. Therefore, he/she loses 1 mark.

Candidate A

(b) The resistivity of a semiconductor increases with temperature ✗. This is because the ions vibrate a lot and electrons make many collisions with the ions ✗.

Candidate B

(b) Increasing the temperature frees more electrons from the ions ✓. This leads to a decrease in resistivity when the temperature increases ✓.

Candidate A has given the variation of resistivity for a *metal*, not for a semiconductor. The candidate has not answered the question and therefore no marks are awarded. Even A-grade candidates can lose marks by not reading the question properly or by rushing through an answer. In contrast, Candidate B has the correct physics and secures 2 marks.

Candidate A

(c) (i) area = $\pi \times (0.34 \times 10^{-3})^2 = 3.63 \times 10^{-7}\,\text{m}^2$ ✓

$R = \dfrac{\rho L}{A} = \dfrac{7.8\times10^{-6}\times8.5\times10^{-2}}{3.63\times10^{-7}}$ ✓

$R = 1.83\,\text{ohm}$ ✓

Candidate B

(c) (i) resistance = $\dfrac{7.8\times10^{-6}\times0.085}{\pi\times(0.00034)^2} = 1.8$ ✓ ✓ ✗

Candidate A has produced a well-structured answer and scores 3 marks. He/she has worked in SI units and written the data in standard form. Candidate B has quoted the answer correctly to two significant figures but has forgotten to include the unit. Consequently, Candidate B scores only 2 marks.

Candidate A

(c) (ii) current = $\dfrac{V}{R} = \dfrac{12}{1.83+0.50}$ ✓ ✓

current = $5.15 \approx 5.2\,\text{A}$ ✓

Candidate B

(c) (ii) $V = IR$ ✓

$12 = I \times 1.8$ ✗

current = 6.7 amps ✗

Candidate A has included the internal resistance of the supply in the calculation. The examiner will have no problem in following this response. Candidate B scores 1 mark for the correct equation. However, the candidate has not answered the question because the internal resistance of the supply has been ignored. The resistance of the circuit is not 1.8 Ω — it is 2.3 Ω.

Overall, Candidate A scores 8 out of 10 marks, which is borderline grade A. Candidate B scores 6 marks. He/she does not have a good understanding of basic circuits in series, which cost 2 marks in (c)(ii). Missing out the unit in (c)(i) is almost unforgivable at AS. Candidate B's marks could have been improved by careful scrutiny of the questions and by paying more attention to detail.

Question 4

Power; series and parallel circuits

(a) State Kirchhoff's second law. (1 mark)

(b) You are provided with three identical cells. Each has an e.m.f. of 1.5 V. The cells are connected in series to an electrical circuit. State, and explain, the minimum e.m.f. that can be provided by the three cells. (2 marks)

(c) The diagram below shows an electrical circuit.

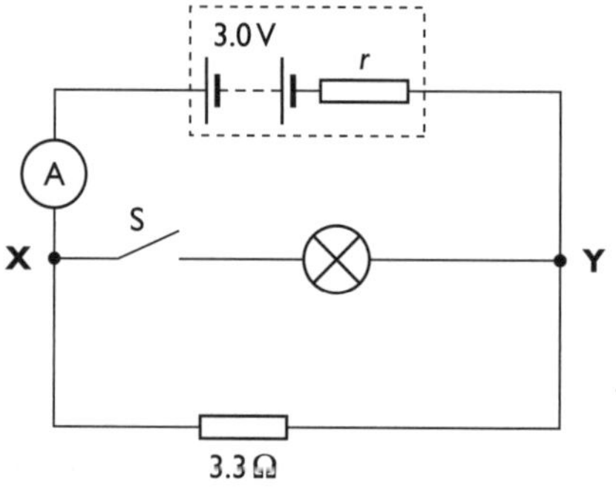

The battery has an e.m.f. of 3.0 V and internal resistance *r*. The p.d. across the filament lamp is 2.5 V and its power rating at this voltage is 0.75 W.

(i) Describe what happens to the ammeter reading when the switch, S, is closed. (2 marks)

(ii) Determine the resistance of the lamp. (2 marks)

(iii) Determine the total resistance between X and Y. (2 marks)

(iv) Determine the internal resistance *r* of the cell. (3 marks)

Total: 12 marks

Candidates' answers to Question 4

Candidate A

(a) The sum of e.m.f.s in a loop = the sum of p.d.s in the same loop ✓.

Candidate B

(a) The total e.m.f. in a loop of a circuit is equal to the total voltages in the same loop ✓.

e Both candidates have made a good start. Remember that the 'voltage' is the same as potential difference (p.d.).

Candidate A

(b) The minimum e.m.f. is 1.50 V ✓. This is shown by the circuit:

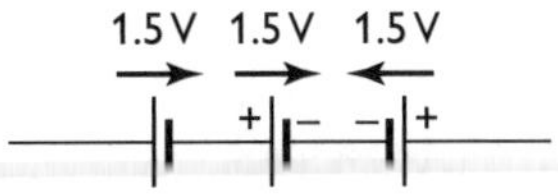

✓

This makes the e.m.f. equal to (1.5 V + 1.5 V – 1.5 V) = 1.5 V

Candidate B

(b) I would connect the cells in series with one having opposite polarity ✓.

e Candidate A has been sensible and provided a circuit diagram. Good physics has been demonstrated by showing that the e.m.f. of one of the cells has to be subtracted. Candidate B loses 1 mark for not stating the value of the e.m.f.

Candidate A

(c) (i) The ammeter reading will go up (increase) ✓. This is because the total resistance between X and Y decreases ✓.

Candidate B

(c) (i) The resistance between X and Y is 3.3 Ω when S is open. When S is closed, the current increases. The ammeter reading increases ✓.

e Candidate A has provided a complete answer by saying what happens to the current and giving the reason for the increase in the current. Candidate B has not given any reason for the change in the ammeter reading and, therefore, loses 1 mark.

Candidate A

(c) (ii) $P = \frac{V^2}{R}$

$R = \frac{2.5^2}{0.75} = 8.33\,\Omega$ ✓ ✓

Candidate B

(c) (ii) current in lamp $= \frac{P}{V} = \frac{0.75}{2.5} = 0.30\,\text{A}$ ✓

resistance $= \frac{V}{I} = \frac{2.5}{0.30} = 8.33\,\Omega$ ✓

e The candidates have used different approaches successfully and each is awarded the 2 marks. The answers provided by both candidates are well structured.

Candidate A

(c) (iii) $\frac{1}{R} = \frac{1}{3.3} + \frac{1}{8.33} = 0.423$ ✓

resistance $= \frac{1}{0.423} = 2.36\,\Omega$ ✓

Candidate B

(c) (iii) $R = \dfrac{3.3 \times 8.33}{3.3 + 8.33} = 2.36\,\Omega$ ✓ ✓

Once again, both candidates have done well and picked up maximum marks.

Candidate A

(c) (iv) p.d. across internal resistor = 3.0 − 2.5 = 0.5 V ✓

$\text{current} = \dfrac{2.5}{2.36} = 1.06\text{ A}$ ✓

$\text{resistance} = \dfrac{V}{I} = \dfrac{0.5}{1.06} = 0.47\,\Omega$ ✓

Candidate B

(c) (iv) $E = I(R + r)$ ✓

Not sure what to do here ✗.

The logical steps of Candidate A's response indicate an excellent understanding of circuits. Candidate B has been daunted by this question, writing an equation but failing to substitute any values. The examiner has been kind and has awarded him/her 1 mark.

Overall, Candidate A scores the maximum 12 marks. The answers are brief and contain just the right amount of mathematics. Candidate B scores 8 marks, half of these from (c)(ii) and (c)(iii). The last part of the question has defeated Candidate B, who was unable to finish the calculation for the internal resistance of the battery. There will always be some challenging questions on the examination paper.

Question 5

Practical circuits

The circuit diagram below shows a simple electrical circuit used to monitor temperature changes in a room.

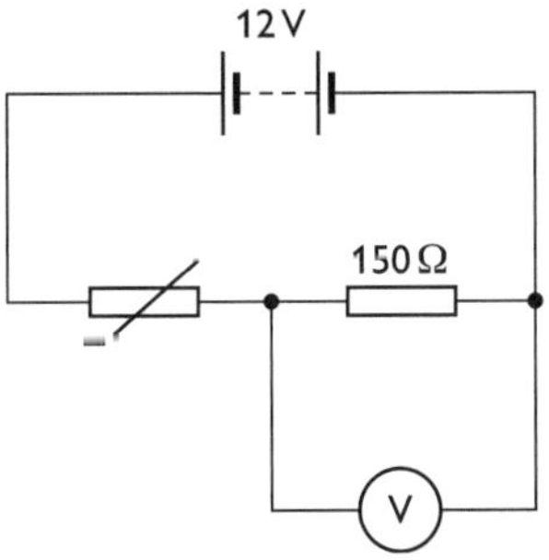

The battery has e.m.f. 12 V and negligible internal resistance. The voltmeter placed across the resistor of resistance 150 W has an infinite resistance. With the NTC thermistor at room temperature, the current in the circuit is 30 mA.

(a) Determine the resistance of the thermistor at room temperature. (3 marks)

(b) The temperature of the thermistor is increased. State, and explain, the effect of this increase on:

(i) the current in the circuit (3 marks)

(ii) the voltmeter reading (2 marks)

Total: 8 marks

Candidates' answers to Question 5

Candidate A

(a) p.d. across the resistor = $30 \times 10^{-3} \times 150 = 4.5\,\text{V}$ ✓

I can find the resistance R of the thermistor using the potential divider equation:

$$4.5 = \frac{150}{150 + R} \times 12 \checkmark$$

$$R = \frac{150 \times 12}{4.5} = 400\,\Omega \;✗$$

Candidate B

(a) The p.d. across the resistor is $V = IR = 0.030 \times 150 = 4.5\,\text{V}$ ✓

The p.d. across the thermistor is $V = 12 - 4.5 = 7.5\,\text{V}$ ✓

$$\text{resistance} = \frac{V}{I} = \frac{7.5}{0.030} = 250 \text{ ohms} \checkmark$$

Candidate B's answer is clearly set out and indicates good knowledge of basic circuits. Candidate A uses the potential divider equation, but makes a mistake in the final step by determining the total resistance of the circuit. The resistance of the thermistor is found by subtracting 150 Ω from 400 Ω.

Candidate A

(b) (i) The resistance of the thermistor will decrease ✓. This is because there are now more free electrons in the circuit. The total resistance of the circuit will decrease ✓. According to $I = V/R$, the current is inversely proportional to the resistance. Therefore, the current will increase ✓.

Candidate B

(b) (i) The current will increase ✓. This is because resistance of the thermistor has decreased ✓.

Both candidates have done well. Candidate A scores 3 marks and Candidate B scores 2. Candidate A's response is precise with a good explanation of the cause of the decrease in the current. Candidate has been too brief by offering only two statements for a question worth 3 marks.

Candidate A

(b) (ii) I can use the potential divider equation. As the resistance of the thermistor decreases it takes a smaller share of the supply voltage ✓. Hence the p.d. across the resistor (voltmeter reading) increases ✓.

I have used the equation $\frac{V_1}{V_2} = \frac{R_1}{R_2}$.

Candidate B

The voltmeter reading will decrease because the current has increased ✗.

Candidate A has provided a comprehensive answer and continues to demonstrate good knowledge of circuits. Candidate B has probably made a guess.

Overall, Candidate A scores 7 marks and remains on target for a grade A. The comprehensive answers given by this candidate show an excellent knowledge of circuits. Candidate B scores 5 marks. Picking up 1 or 2 more marks would have boosted the grade of this candidate. Candidate B displays the typical symptoms of a grade-C candidate who has a shallow understanding of some topics.

Question 6

Wave motion

(a) Show that the speed v of a wave is related to the wavelength λ and frequency f by the wave equation, $v = f\lambda$. (2 marks)

(b) A loudspeaker is connected to a signal generator. The graph below shows the displacement s against time t graph for an air particle at a certain distance from the loudspeaker.

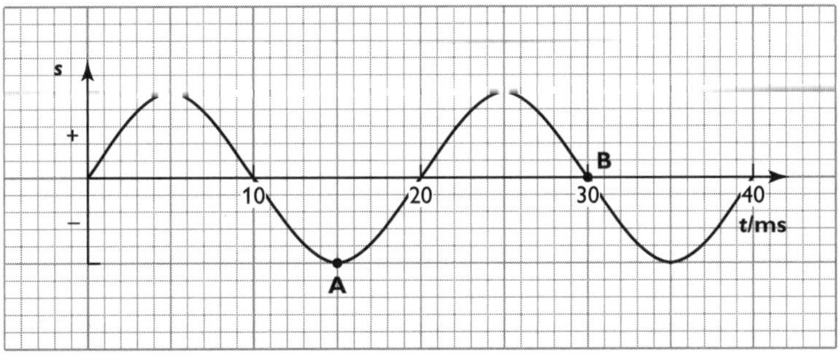

The loudspeaker emits sound waves of power 8.0 mW. The loudspeaker cone has a cross-sectional area of $5.0 \times 10^{-3}\,m^2$.

(i) Determine the frequency of the sound. (2 marks)

(ii) Determine the phase difference between A and B. (2 marks)

(iii) Calculate the intensity of the sound emitted by the loudspeaker. (2 marks)

(iv) State, and explain, any changes that there would be in the displacement against time graph when the loudspeaker emits sound of the same frequency but the power emitted from it is doubled. (2 marks)

Total: 10 marks

Candidates' answers to Question 6

Candidate A

(a) speed = distance/time

In a time of 1 s, there are f wavelengths produced ✓. The distance travelled by the waves in 1 second is $f \times \lambda$ ✓. The distance travelled per second is the wave speed, v. Hence, $v = f\lambda$.

Candidate B

(a) In one oscillation, the waves move forward a distance of wavelength λ. In a time of one period, T, the distance travelled by the wave is λ ✓.

speed = $\dfrac{\lambda}{T} = \lambda \times \left(\dfrac{1}{T}\right) = \lambda \times f$ ✓

The candidates have answered the question differently but both responses are acceptable, for 2 marks. This is a good start from both candidates.

Candidate A

(b) (i) period = 20 ms = 20×10^{-3} s ✓

frequency = $\dfrac{1}{20 \times 10^{-3}}$ = 50 Hz ✓

Candidate B

(b) (i) frequency = $\dfrac{1}{20 \times 10^{-3}}$ = 50 hertz ✓ ✓

Both candidates have done well and have picked up 2 easy marks.

Candidate A

(b) (ii) The points are separated by $\frac{3}{4}$ of a period ✓. There is a phase difference of 270° between the two points ✓.

Candidate B

(b) (ii) The points do not vibrate in phase ✗. The time between A and B is 75% of a period, *T* ✓.

Candidate A's response is well structured, for 2 marks. Candidate B has not provided a numerical answer for the phase difference. The candidate earns 1 mark for realising that points A and B are separated by $\frac{3}{4}$ of a period.

Candidate A

(b) (iii) intensity = $\dfrac{\text{power}}{\text{area}}$

$= \dfrac{8.0 \times 10^{-3}}{5.0 \times 10^{-3}} = 1.6\ \text{W m}^{-2}$ ✓ ✓

Candidate B

(b) (iii) intensity = $\dfrac{8.0}{0.005}$ = 1600 ✓ ✗

Candidate A has again provided a perfect answer, earning both marks. Candidate B has not changed the power into watts (W) and not given the units. The examiner has awarded a generous mark for substituting the quantities into a correctly identified equation.

Candidate A

(b) (iv) power $\propto$ intensity of sound

intensity $\propto$ amplitude^2 ✓

The intensity is doubled; hence the amplitude increases by a factor of $\sqrt{2}$ ✓.

Candidate B

(b) (iv) The amplitude will go up because of more power from the speaker ✗.

This is a tough question, targeted at grade-A/B candidates. It is quantitative, so examiners expect an exact answer for the amplitude. Candidate A has done well, demonstrating a good understanding of intensity and power, which are difficult concepts. Candidate B has been too brief and has not understood that the question requires an analytical response.

Overall, Candidate A scores the full 10 marks and typifies the capabilities of a high-scoring candidate. The answers contain a good blend of physics and mathematics. The knowledge base of this candidate is good enough to achieve a grade A. Candidate B scores 6 marks. An extra mark could have gained in (b)(iii) by including the unit for the intensity, mW m^{-2} (milliwatts per square metre). Careless work has cost this candidate at least one grade.

Question 7

Electromagnetic waves

(a) State two main properties of electromagnetic waves. (2 marks)

(b) Describe some of the harmful and beneficial effects of ultraviolet radiation. (3 marks)

(c) State a typical wavelength of radio waves. Use your value for the wavelength to estimate the frequency in kHz of radio waves. (3 marks)

(d) Light reflected from a shiny surface, such as water or glass, is partially plane polarised.

(i) Explain what is meant by plane-polarised light. (1 mark)

(ii) Describe how you could demonstrate that reflected light is partially polarised. (2 marks)

Total: 11 marks

Candidates' answers to Question 7

Candidate A

(a) All electromagnetic waves can travel through a vacuum ✓. They are also all transverse waves with different frequencies ✓. All electromagnetic waves inter act as photons.

Candidate B

(a) These waves can travel through space where there is a vacuum ✓. EM waves travel at a speed of 300 000 000 $m\,s^{-1}$ in a vacuum ✓.

e Candidate A has offered three correct statements for a part question worth 2 marks. Examiners cannot award bonus marks, but perhaps this candidate wanted to make sure that both marks were secured. Candidate B has also done well by recalling two main properties. This is a good start from both candidates.

Candidate A

(b) UV-A and UV-B can cause wrinkling of the skin and cancer respectively ✓. UV-B is also good because it makes vitamin D in our bodies ✓ ✓.

Candidate B

(b) Ultraviolet radiation can cause cancer ✓. One type of ultraviolet makes vitamin D in our skin ✓.

e Candidate A has named the different types of UV radiation and correctly identified UV-B as the radiation that produces vitamin D. Candidate B loses 1 mark for not identifying the type of UV responsible for producing vitamin D.

Candidate A

(c) 1.0 km ✓

$c = f\lambda$

$f = \frac{c}{\lambda} = \frac{3.0 \times 10^8}{1.0 \times 10^3} = 3.0 \times 10^5\,\text{Hz} = 300\,\text{kHz}$ ✓ ✓

Candidate B

(c) frequency = $\frac{300\,000\,000}{200} = 1\,500\,000\,\text{Hz}$ ✓ ✓ ✗

Candidate A has done well. The wavelength and the physics (wave equation) used to get the answer are stated clearly. Candidate B has the wavelength embedded in the substitution into the wave equation. The answer is correct, but it is not in kilohertz, so a mark is lost.

Candidate A

(d) (i) The electric field oscillates at right angles to the velocity and is in just one plane ✓.

Candidate B

(d) (i) The vibrations are in one plane only. The wave still vibrates at 90° to the direction in which the light travels ✓.

Both candidates earn the mark. The key marking point was vibrations or oscillations in 'one plane'.

Candidate A

(d) (ii) I would use a polarising filter known as a Polaroid ✓. I would point the Polaroid in the direction of the light and rotate it until the transmitted light has minimum intensity. The light is polarised if rotating the Polaroid by a further by 90° makes the intensity of light maximum ✓. Further rotation of the Polaroid by 90° will give minimum intensity again.

Candidate B

(d) (ii) The light must also have vibrations in other planes ✗.

The answer from Candidate A is perfect and shows an excellent understanding of polarisation. Candidate B has not answered the question at all because there is no mention of a polarising filter.

Overall, Candidate A scores all 11 marks. The knowledge of this candidate is clear to see from the way the answers are represented. Candidate B gains 7 marks. Losing 1 mark in (c) for not giving the answer in kHz was careless. He/she also has only a superficial knowledge of polarisation.

Question 8

Interference

(a) State what is meant by a progressive wave. (1 mark)

(b) Two microwave transmitters produce waves of the same frequency. State two conditions that must be satisfied for the waves from the two transmitters to produce an interference pattern. (2 marks)

(c) A parallel beam of monochromatic light of wavelength 630 nm is incident normally at a diffraction grating. The diffraction grating has 200 lines per millimetre.

(i) Calculate the separation, *d*, in metres between adjacent lines on the grating (this is also known as the grating spacing or the grating element). (1 mark)

(ii) Calculate the angle, θ, between the first-order maximum and the second-order maximum. (3 marks)

(iii) Determine the total number of bright spots that would be seen beyond the diffraction grating. (3 marks)

Total: 10 marks

Candidates' answers to Question 8

Candidate A

(a) A progressive wave is also known as a travelling wave. It carries energy between places through vibrations ✓.

Candidate B

(a) This is a wave that transports energy as it vibrates ✓.

e Candidate A's response is perfect and demonstrates a clear understanding of progressive waves. Candidate B has given an adequate answer and also earns the mark.

Candidate A

(b) The waves must meet and have similar amplitudes ✓. The transmitters must emit coherent waves ✓.

Candidate B

(b) The waves from the transmitters must have the same wavelength and frequency ✗. They must also have the same amplitude ✓.

e Candidate A has again produced a model answer. Candidate B is simply repeating the information given in the question and has missed out the key idea that the waves must show coherence.

Candidate A

(c) (i) The spacing between the lines is d.

$$d = \frac{1\times10^{-3}}{200} = 5.0 \times 10^{-6}\,\text{m}\ ✓$$

Candidate B

(c) (i) The number of lines in one metre will be 200 × 1000 = 200 000 lines

Therefore $d = \dfrac{1\,\text{m}}{200\,000\text{ lines}} = 0.000\,005\,\text{m}$ ✓

The candidates have used different approaches but both obtained the same answer. Candidate A's response is preferable because it is concise and the answer is expressed in standard form.

Candidate A

(c) (ii) $d\sin\theta = n\lambda$ ✓

$d = 5.0 \times 10^{-6}\,\text{m}$; $\lambda = 630\,\text{nm} = 630 \times 10^{-9}\,\text{m}$; $n = 2$

Therefore:

$$\sin\theta = \frac{n\lambda}{d} = \frac{2\times630\times10^{-9}}{5.00\times10^{-6}} = 0.252\ ✓$$

$$\theta = \sin^{-1}(0.252) \approx 14.6°\ ✓$$

Candidate B

(c) (ii) $d\sin\theta = 2\lambda$ ✓

$$\sin\theta = 2 \times \frac{0.000\,000\,000\,6.3}{0.000\,005} = 0.252\ ✓\ ✗$$

The structure of Candidate A's answer is excellent. Candidate B has not used standard form and this makes the answer look messy. The calculation is not finished — only the sine of the angle has been determined and not the angle, θ.

Candidate A

(c) (iii) The maximum value for the angle θ is 90° and $\sin 90° = 1$ ✓.

$d\sin\theta = n\lambda$ becomes $d = n_{max}\lambda$

The maximum number of orders $n_{max} = \dfrac{d}{\lambda} = \dfrac{5.0\times10^{-6}}{630\times10^{-9}} = 7.94 = 7$ (integer needed) ✓

There are 14 bright spots ✗.

Candidate B

(c) (iii) The maximum angle must be 90 degrees ✓. There must be 100 orders ✗.

Candidate A lost the final mark because there are 15 orders — the zeroth order and seven spots on either side of the zero direction. Candidate B managed to secure 1 mark by stating that the maximum angle must be 90°. The answer '100 orders' is a guess.

Overall, Candidate A scores 9 out of 10 marks and is on target to obtain an A grade. Candidate B has an adequate understanding of this topic but has lost 4 valuable marks for careless or incomplete answers. Not calculating the angle in (c)(ii) has cost this candidate at least one grade. He/she also needs to scrutinise the questions before answering.

Question 9

Stationary waves

(a) State one similarity and one difference between a stationary wave and a progressive wave. (2 marks)

(b) A stretched string is fixed at each end. It is plucked at the midpoint of its length.

(i) Explain how a stationary wave is produced in this string. (3 marks)

(ii) Sketch the shape of the stationary-wave pattern. (2 marks)

(c) A tube is open at both ends. A loudspeaker placed at one of the open ends produces a loud resonating sound corresponding to a fundamental note. A stationary wave with one node is produced in the air column of the tube. The length of the tube is 15 cm. Determine the frequency of the sound emitted from the loudspeaker. (The speed of sound in air is 340 m s^{-1}.) (3 marks)

Total: 10 marks

Candidates' answers to Question 9

Candidate A

(a) Similarity: both types of waves vibrate ✓.
Difference: a progressive wave transfers energy between two points whereas a standing (stationary) wave does not ✓.

Candidate B

(a) Stationary waves have zero speed. They cannot transfer energy ✓.
Both stationary and progressive waves are waves ✗.

e Candidate A has made an excellent start. Both candidates gain 1 mark for appreciating that a stationary wave does not transfer energy. The similarity given by Candidate B is not an adequate answer.

Candidate A

(b) (i) The progressive waves on the string travel in opposite directions. The waves are reflected at the fixed ends ✓. This produces two waves travelling in opposite directions ✓ that combine in accordance with the principle of superposition ✓.

Candidate B

(b) (i) A stationary wave is created when two waves travelling in opposite directions combine ✓.

e Candidate A gains all 3 marks. Candidates B has written only one sentence in answer to a part question worth 3 marks. The award of 1 mark is kind because Candidate B has not given an answer that is specific to the question.

Candidate A

(b) (ii) The standing wave pattern has a node at each of the fixed ends and one antinode in the middle.

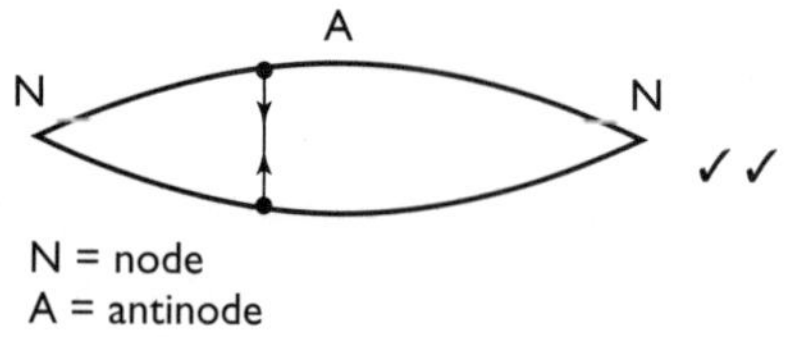

Candidate B

(b) (ii) The pattern is:

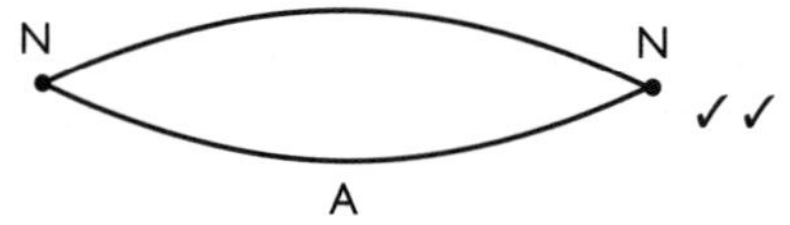

Both candidates have produced good answers. Candidate A has included a description that further supports the sketch. If you have time in the examination, this is always a good idea.

Candidate A

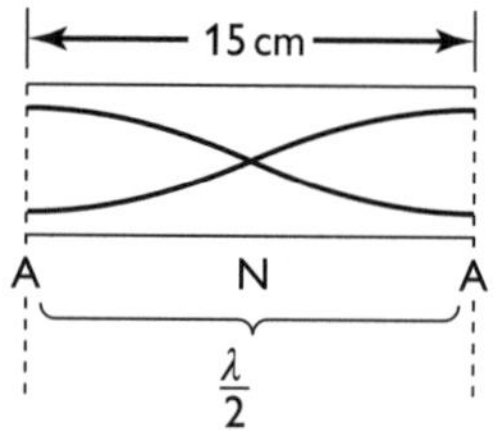

(c) There are antinodes at the ends of the tube and one node in the middle.

$\frac{\lambda}{2} = 0.15$

wavelength, $\lambda = 0.30\,\text{m}$ ✓

$v = f\lambda$

$f = \frac{v}{\lambda} = \frac{340}{0.30}$ ✓

frequency = 1.13 kHz ✓

Candidate B

(c) The wavelength is twice the length of the tube because the separation between two antinodes is half the wavelength.

wavelength = 30 cm ✓

frequency = $\frac{340}{30}$ = 11.3 Hz ✓ ✗

Candidate A benefits from drawing a sketch of the stationary wave pattern. The answer is once again well structured. Candidate B has lost an easy mark by not converting the wavelength into metres.

Candidate A scores the full 10 marks and Candidate B scores 6 marks. Picking up a couple of marks would have helped Candidate B to secure a grade higher than C. This candidate should have realised that the 3 marks available in (b)(i) indicate that a detailed answer is required. Not converting the wavelength into metres in (c) has also proved costly.

Question 10

Photons and the photoelectric effect

The work function of magnesium is 3.7 eV. A metal plate coated with magnesium is charged negatively by *momentarily* connecting it to a DC power supply. The plate is then illuminated with electromagnetic waves of wavelength 3.2×10^{-7} m. Photoelectrons are emitted from the surface of the plate.

(a) Explain what is meant by the term 'work function'. (1 mark)

(b) Describe how the incident electromagnetic radiation interacts with the magnesium to release the photoelectrons. (4 marks)

(c) Calculate the maximum kinetic energy, in joules, of the photoelectrons emitted from the plate. (4 marks)

Total: 9 marks

Candidates' answers to Question 10

Candidate A

(a) The work function of a metal is the minimum energy needed to free an electron from the surface of the metal ✓.

Candidate B

(a) This is the minimum energy an electron needs to free itself from the metal ions ✓.

Both candidates have correctly answered this opening question.

Candidate A

(b) The EM waves behave like 'particles', called photons ✓. A single photon interacts with a single electron ✓. The electron is released when the photon has energy greater than the work function energy ✓. Energy is conserved between the photon and the electron ✓. The incident waves must have frequency greater than (or equal to) the threshold frequency.

Candidate B

(b) A photon gives its energy to a single electron. It is a 'one-to-one' interaction ✓. Energy is conserved when a photon gives its energy to the electron ✓. Einstein's photoelectric equation applies: $hf = \phi$ + maximum kinetic energy of electron ✓.

Candidate A has provided more than four correct statements but the examiner can only award a maximum of 4 marks. The statements are clear and expressed in appropriate scientific terms. Candidate B has also done well but has given only three statements. These are correct, so he/she scores 3 marks.

Candidate A

(c) $\frac{hc}{\lambda} = \phi + KE_{max}$

$\phi = 3.7 \times 1.6 \times 10^{-19} = 5.92 \times 10^{-19}$ J ✓

energy of photon $= \frac{6.63 \times 10^{-34} \times 3.0 \times 10^{8}}{3.2 \times 10^{-7}} = 6.22 \times 10^{-19}$ J ✓ ✓

$KE_{max} = (6.22 - 5.92) \times 10^{-19} = 3.0 \times 10^{-20}$ J ✓

Candidate B

(c) $E = \frac{6.63 \times 10^{-34} \times 3.0 \times 10^{8}}{3.2 \times 10^{-7}} \approx 6.2 \times 10^{-19}$ joules ✓ ✓

$hf = \phi$ + maximum kinetic energy of electron

$\phi = 3.7$ J ✗

maximum kinetic energy = 6.2 − 3.7 = 2.5 J ✗

Candidate A has made good use of the photoelectric equation. Instead of substituting all the values into the equation, this candidate has determined logically the maximum kinetic energy of the photoelectrons. Candidate B gains 2 marks for the photon energy. Serious errors with powers of ten have prevented this candidate from picking up any further marks and the work function has not been converted from electronvolts to joules.

Overall, Candidate A scores the full 9 marks and is on target for an A grade. Candidate B scores 6 marks and has done well on this tough question. However, the response indicates gaps in his/her knowledge and understanding of the photoelectric effect.

Question 11

Wave–particle duality

(a) Write the de Broglie equation and define all the terms. (2 marks)
(b) Slow-moving neutrons from a nuclear reactor have a speed of 1.2 km s^{-1}. Calculate the de Broglie wavelength of the neutrons. (2 marks)
(c) Use your answer to (b) to suggest why such neutrons would be suitable in investigating the atomic structure of matter. (1 mark)
(d) Describe how, in the laboratory, you could demonstrate the wave-like behaviour of electrons. (3 marks)
(e) Describe how a graph of the de Broglie wavelength, λ, of an electron against speed^{-1} can be used to determine a value for the Planck constant, *h*. (2 marks)

Total: 10 marks

Candidates' answer to Question 11

Candidate A

(a) $\lambda = \frac{h}{p}$ ✓

where λ is the de Broglie wavelength, h is the Planck constant and p is the momentum of the particle ✓.

Candidate B

(a) $\text{wavelength} = \frac{\text{Planck constant}}{\text{mass of particle} \times \text{speed of particle}}$ ✓✓

Both candidates gain 2 marks by completely different approaches — the word equation written by Candidate B is elegant.

Candidate A

(b) $\lambda = \frac{h}{mv} = \frac{6.63\times10^{-34}}{1.7\times10^{-27}\times1200} = 3.25 \times 10^{-10}\,\text{m}$ ✓✓

Candidate B

(b) $\text{wavelength} = \frac{6.63\times10^{-34}}{1.7\times10^{-27}\times1200} = 3.25 \times 10^{-10}\,\text{m}$ ✓✓

Both candidates score 2 marks.

Candidate A

(c) The wavelength of the neutrons is roughly the same as the separation between atomic layers. The neutrons are, therefore, diffracted by matter ✓.

Candidate B

(c) The neutrons can do this because they have no charge ✗.

e Candidate A has done well to secure this difficult mark. The slow-moving neutrons behave like a wave. They are diffracted because the de Broglie wavelength is similar to the dimensions of the atoms and the separation between them, which is approximately 10^{-10} m.

Candidate A

(d) Electrons are accelerated in a diffraction tube. They pass through a thin layer of graphite ✓. The electrons have a wavelength given by the de Broglie equation in part (a). The electrons are diffracted by the carbon atoms and the atomic layers of carbon atoms ✓. We can tell that the electrons are diffracted because they produce diffraction rings on the fluorescent screen ✓.

Candidate B

(d) According to de Broglie, electrons travel as waves. They are made to pass through carbon atoms ✓.

e Candidate B has been brief. The mark is earned for identifying that the electrons pass through graphite or carbon atoms. Candidate A has no problems in scoring all 3 marks.

Candidate A

(e) A graph of λ against v^{-1} is a straight line through the origin ✓.

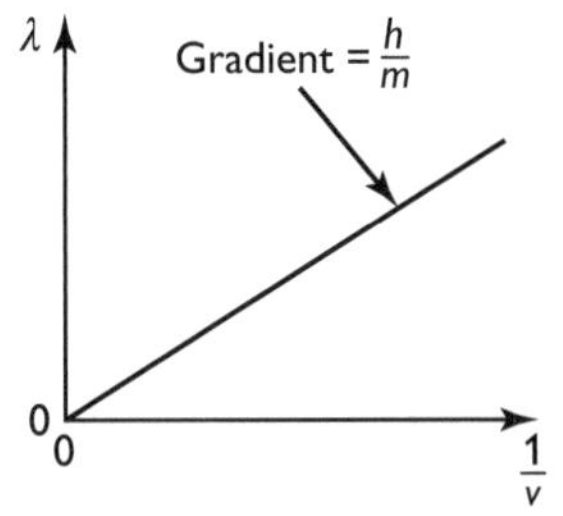

This is because $\lambda = \left(\frac{h}{m}\right) \times \frac{1}{v}$.

The gradient of the line is h/m. Therefore, h = mass of electron × gradient of line ✓.

Candidate B

(e) I think that the graph will be a straight line and mostly likely will pass through the 0, 0 point (the origin) ✓.

e Well done to Candidate A for a flawless answer. The de Broglie equation has been used as a guide to show how the Planck constant can be determined from the gradient of the line. Candidate B has most likely guessed. The examiner awards 1 mark for the idea of a straight line through the origin.

Overall, Candidate A scores the full 10 marks. The understanding of the physics comes through at all stages with this candidate. Candidate B scores 6 marks and will obtain a grade C. Grade-C candidates are often defeated by the complexities of wave–particle duality, so Candidate B has not done too badly here.

Question 12

Energy levels in atoms

(a) Explain what is meant by the term 'energy level'. (1 mark)

(b) Explain why the spectrum of light from the Sun has dark narrow lines. (2 marks)

(c) The diagram below shows some of the energy levels of mercury atoms.

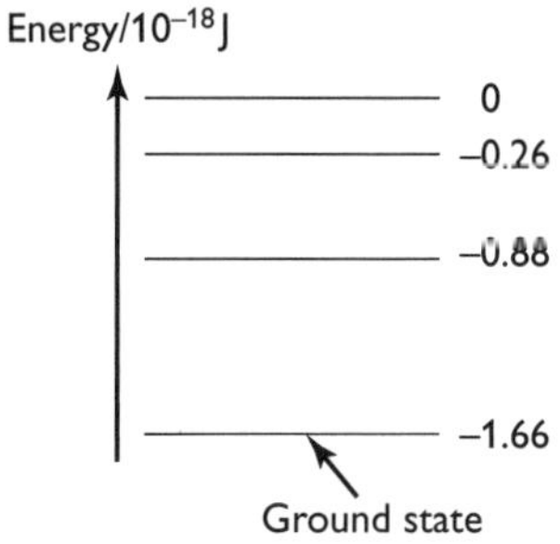

(i) Calculate the wavelength of electromagnetic waves emitted when an electron makes a transition between the energy levels at -0.88×10^{-18} J and -1.66×10^{-18} J. (3 marks)

(ii) A proton with kinetic energy of 0.62×10^{-18} J collides with a mercury atom and loses all its energy. Explain how the atom absorbs this energy. (2 marks)

(iii) Explain whether or not a photon of energy 0.65×10^{-18} J can interact with a mercury atom. (2 marks)

Total: 10 marks

Candidates' answer to Question 12

Candidate A

(a) The energy of an electron inside the atom is quantised. The quantised state is known as an energy level ✓.

Candidate B

(a) An electron can only have certain allowed energies — these are the energy levels ✓.

e Both candidates have done well with this tough opening part question. The answer from Candidate B is uncharacteristic of a grade-C candidate.

Candidate A

(b) The dark lines are absorption spectral lines ✓. These occur because of absorption of certain wavelengths (or photons) by the atoms ✓.

Candidate B

(b) The lines are due to certain wavelengths from the sunlight being absorbed by gas atoms of the Sun ✓.

Both candidates have correctly identified the lines as absorption lines. However, only Candidate A has given an answer of sufficient depth to earn both marks.

Candidate A

(c) (i) $\frac{hc}{\lambda} = \Delta E$ ✓

$\lambda = \frac{hc}{\Delta E} = \frac{6.63 \times 10^{-34} \times 3.0 \times 10^{8}}{(1.66 - 0.88) \times 10^{-18}}$ ✓

wavelength emitted = 2.55×10^{-7} m (255 nm) ✓

Candidate B

(c) (i) energy of photon = $(1.66 - 0.88) \times 10^{-18}$ J ✓

$hf = 0.78 \times 10^{-18}$

$f = 1.18 \times 10^{15}$ Hz ✓

$\lambda = \frac{c}{f} = 2.55 \times 10^{-7}$ m ✓

There are no problems at all with the answer from Candidate A. Candidate B has also gained all 3 marks but has not shown all the working. On this occasion, the substitution was correct, but this is a risky strategy.

Candidate A

(c) (ii) The energy of the photon is the same as the difference between the energy levels -0.26×10^{-18} J and -0.88×10^{-18} J ✓. An electron absorbs this energy from the photon and makes a transition between these two levels ✓. It increases its energy by making a quantum jump.

Candidate B

(c) (ii) The electron will jump from the −0.88 level to the −0.26 level ✓.

The examiner has been rather generous in awarding 1 mark to Candidate B. The powers of ten are missing from the energy levels, but the examiner has no doubt about which two energy levels are involved in the exchange of the energy between the photon and the electron. Candidate A has once gain provided an perfect answer, for 2 marks.

Candidate A

(c) (iii) The energy of the photon does not match with the difference between any two energy levels ✓. The exchange of energy between the photon and the atom is not possible ✓.

Candidate B

(c) (iii) Photons can always be absorbed by atoms ✗.

Candidate A has a good understanding of energy levels and deserves the 2 marks. Candidate B has been overwhelmed by this question.

Overall, Candidate A scores the full 10 marks — the answers are worthy of a grade A. Candidate B unexpectedly scores 6 marks. Energy levels in atoms is a topic that grade-C candidates find challenging, so Candidate B has done well, particularly with the earlier part questions.